Resolute Productivity

Leveraging Technology to Thrive in the New Workspace

By Gregory Shefler

ISBN 978-1-7374091-3-7 (paperback)

Dedicated to my father, who taught me what it meant to "earn the day".

Contents

Introduction

Time is the school in which we learn,
Time is the fire in which we burn.

— Delmore Schwartz

A life worth living is one of purpose. Life is also short and filled with distractions and stressors that impede our ability to realize our purpose. These facts of life accentuate the gravity and urgency in how we use our time in service of our higher goals. The technological world of the 21st Century has greatly increased our capacity to achieve goals. Unfortunately, this same technology has also burdened us with significant hindrances to human self-realization. It contributed to increased expectations of us in terms of our baseline work output. Sadly, such elevated levels production often came without a proportional increase in both monetary and psychological benefit. Additionally, internet technology has proven to be an unprecedented source of superfluous distraction in every facet life. Such an overabundance of the irrelevant dilutes our ability to see what truly serves our purposes and drives. Without the skills and capacities to navigate these hindrances, it is easy to unwittingly live a life of dispensable achievement and unrealized purpose.

These stark realities of the modern world have given rise to the now-booming productivity industry. Every year, there are new "perfect" systems of productivity and organization popping on. Productivity gurus emerge with impressive regularity - all claiming to have figured out the key to getting things done and creating a fulfilling life. App stores are awash with software and paid services promising to automatically get your life in order. Long-running extant software suites like the venerable Microsoft Office now regularly add new productivity features. One would think that out of these innumerable "solutions", the key to ultimate productivity and organization would be unlocked. If only it were so easy.

A Human Conception of Productivity

The term "productivity" has taken on a larger meaning in recent years. The dictionary lists productivity as merely the effectiveness in which effort creates finished output. This idea of productivity would be worth striving for if we were but mere automatons or work animals. The modern use of the term has a broader scope as it encompasses more than just a job metric. As humans with a psychological and existential need for purpose, our ideal of productivity must be more profound. If you want to live a life worth living, then you must adopt a better definition of productivity.

In the new broader context, productivity is the capacity to get *meaningful* work done effectively. It entails the ability to parse, prioritize, and process information in an orderly manner. To be productive is to weed out superfluous information and distractions as a means of staying committed to your greater purposes and goals. Productivity is a skill of the mind and its ability to be undistracted and focused on the important tasks at hand.

Though it is often considered a separate subject, organization is a key component of productivity. They are so closely intertwined that productivity cannot exist without organization. Organization is the ability to sort information and ideas concisely and effectively. Productivity cannot exist without organization. For this reason, when the term "productivity" is used in this book, understand that I mean productivity and organization together. In fact, I consider the two terms to be nearly interchangeable.

The Allure of Perfection

The popularity of productivity systems and their proselytizers has fed into a false sense that a perfect system of productivity can exist. It is an alluring myth that there is a single "best way" to manage life and work. Many people have thus gotten caught up in a sense of loyalty and mania about a preferred method of productivity. Still others find themselves on an ultimately fruitless quest to sample every system that exists in the pursuit of the "one". These individuals spend more time in the pursuit of the best productivity system than they do actual work towards achieving their goals.

The allure of perfection has also fostered toxic elements to "productivity culture". It places unrealistic ideations on what being productive and organized should be like. Anything less than being always-on, 100% committed to getting things accomplished indicates that you are lazy and a failure. Many productivity gurus and fanatics boast about needing very little sleep or being extreme early risers. Such habits are worn on their sleeves like badges of honor. Shame is often conferred to those who do not adhere to these and other spartan practices. If this isn't your first productivity book, you probably have experienced at least one author who seemingly wrote their book specifically to brag about their impossibly prolific lifestyle.

Artificial Systems

In truth, all systems of productivity are artificial and imperfect. No one philosophy or approach can address every need of every person. In addition, the nuances and complexities of life are simply too numerous for one system to be able to accommodate them all. The sooner you come to terms with this fact, the better. It does not mean that the pursuit of a productivity system or even the associated skills is pointless. Nearly every productivity system, service, and expert has something useful to offer. It falls to you to find the elements that work best for you and your needs and adapt them as best you can. Inevitably, many of these concepts and approaches will prove to be inadequate. This is perfectly normal, and it only means that you will have to adapt and find better solutions to your changing needs.

Productivity systems can be likened to martial arts styles. Every system of martial arts was created by a founder who noticed that certain fighting techniques and strategies worked best for them and their combat needs. In time, they codified these techniques so they could be taught to others and so a new martial art was born. There is a tremendous diversity in martial arts styles. Every one of them fits the needs of some people and circumstances, but never all of them. Expert martial artists know this and make no claims about the existence of one all-encompassing perfect method of combat. Just as a martial artist must find the system that works best for them, so must you in your quest for your optimal system of productivity.

Regardless of which approach(es) you employ, no system will automatically illuminate the way toward your deeper psychological needs. There is no systemic path toward mental fulfillment or a good work-life balance. These matters are strictly up to you. You must set your goals and maintain the boundaries that facilitate the type of life you want to live. It falls to you to prioritize your life and decide what is important to you. Systems are only as good as the inputs you put into them. If you focus on inconsequential priorities, then you will produce only inconsequential results. Always be mindful of this fact and understand that no system is a panacea or a magical solution to life's challenges.

The Intent of This Book

I believe that the way of productivity is comprised of an interdependent trifecta of three components: Processes, Tools, and Behaviors. These are the indispensable aspects that must comprise an effective system of productivity and organization. Processes are the procedures and routines you follow in sorting information and inputs. Tools are the assortment of software, hardware, and services by which you perform said processes. Behavior is comprised of your habits, mindset, environment, and body that ultimately enable you to utilize these processes.

This book is intended to foster *awareness* of how productivity and organization works in the digital age. I share much of how my personal systems are structured and how they work for me. These systems and sensibilities have served me very well – especially with my easily-distracted, anxious, and depression-prone mind. Anecdotally, I know these systems work. However, I also know that they are entirely subjective and not a good fit for everyone - perhaps not even most people. There is a great diversity and breadth of the topics and concepts covered throughout this book. I hope you come away from this book with a better idea on how broad and all-encompassing productivity can be. It is my ultimate intention that you gain a new or revitalized productive mindset – a step along your journey in finding a way of productivity that helps you achieve your ultimate goals.

Good Luck!

PART I
The New Essential Life Skill

Chapter 1: The Future is Now

The advance of internet technology has changed the way we look at the concept of productivity forever. The core notion of the term simply regards the efficiency in which work turns inputs into output. To increase productivity is to minimize the time it takes to process inputs into outputs. Traditionally, this view of productivity is the same as it always was. However, the coming of the internet and portable tech introduced a new wrinkle in how we approach work. It fostered an unprecedented increase in the rate at which new work inputs reach our desks – all thanks to instant and perpetual connectivity with peers, clients, and information. Knowledge workers found themselves presented with more work than previously thought possible. The same tech that brought more work to the office also provided the requisite tools to be able to process such tasks.

As wonderous as the internet revolution was, it set a new conundrum for knowledge workers: if you cannot learn to properly leverage this new technology, you will be swept away by information overload. The ability to manage abundant emails, appointments, text conversations, file transfers, etc. became vital job skills within only a few short years. Workers who could navigate the information explosion of the internet thrived – those who could not quickly found themselves underemployed or worse. You either learned how to utilize technology or you got left behind.

Just as the dust had settled from the impact the internet made on the workplace, the smartphone revolution happened. From then on, workers could be working at all times of the day just about anywhere in the world. For many people, the divide between work life and personal life had all but disappeared.

The Changing Nature of Work

The trend toward a faster-paced, always-on workspace has continued to increase. The coming of the COVID-19 pandemic further pushed knowledge work into the technological space. Remote work, once thought to be a fanciful dream, proved itself to be a viable and effective way to operate for many companies and workers. Before COVID, many managers regarded remote work as ways in which workers would shirk responsibilities. Much the opposite happened. Where once it was thought remote workers were working from home, it turned out they were living at work! Food delivery, online groceries, and other services exploded during this time. The result was a newfound infrastructure supporting remote workers more than ever before. Remote work also made it harder for unproductive workers to "fake it" by looking busy in front of managers and co-workers. More than ever, results matter.

The waning (or normalization) of the pandemic resulted in a completely transformed notion of what knowledge workspaces could be like. Rather than see organizations return to exclusively on-site operations, many adopted hybrid models. The pandemic proved that remote work was viable and effective. This was in no small part due to the state of internet technology and how it evolved to accommodate remote workforces. Today, the continued evolution of work will see further integration of on-site and remote workforces. In the coming years, these hybrid workplaces will bolster the rise of asynchronous workflows and communication.

One unfortunate change to modern knowledge work is the increased intrusion of work life into personal life. Gone are the days when work could only be done on-site. Cloud technology has made it possible to access workspaces and resources from nearly any device with an internet connection. There was once a time not long ago where the idea of accessing a spreadsheet while on a group phone call with your coworkers from the lobby of a family restaurant – on a single handheld device - was unthinkable. Today, scenarios like this are not unusual and continue to be normalized.

The New Essential Life Skill

With the coming of the post-COVID workspace comes renewed importance on certain life skills. As the new normal continues to set in, these skills are sure to mark the difference between those who thrive or struggle in the coming years. These skills are fourfold. The first of these is the ability to manage an ever-increasing volume of information our work and personal lives generate every day. The second is being able to self-organize and self-motivate to process these inputs without being micro-managed by others. The third skill is having sufficient comfort with technology to be able to leverage it effectively without becoming lost. Finally, there is a renewed need to be able to maintain the boundaries necessary for a work-life balance – something that employers certainly will not do for you.

These four essential skills are all components of a larger overall skillset known as personal productivity. To be able to create and manage a personal productivity system is to be able to thrive professionally and personally. As technology increases the amount of information, tasks, and other inputs coming into our lives, it also provides solutions for effectively managing them. A personal productivity system is the key to thriving in the new workspace.

Conceptually, the flow of modern knowledge work is not much different than it has been in prior decades. Work always entailed processing various types of inputs such as communication, research, scheduling, etc. The key difference today has to do with volume, as technology has created an unprecedented increase in the amounts of inputs, responsibilities, and time constraints. Though technology has increased the volume of work, the way it gets organized and executed has largely remained the same. These same technological advances also hold the key to managing the modern influx of knowledge work. If you know what you are doing, you can leverage these tools in service of an effective productivity system. Done correctly, you can not only manage the modern workflow, but thrive professionally and with a rewarding work-life balance. The first step in this process is to understand what the components of the knowledge workspace are – in both the classic analogue aspect and their modern technological counterparts.

The Analogue Workspace of Yore

The components of a workspace comprise four general categories: the workspace itself, storage, scheduling, and communication. The functions of these spaces are not new and have been part of the workspace since before computers. Every one of these functionalities has a pre- and post-internet counterpart. It is important to recognize this fact, especially when faced with any shiny new technological tool. When evaluating new tech, look to which "classic" analogue role it is designed to supplant. If it does no such thing, then perhaps its value as a productivity tool should be questioned.

The first category within the workspace is…the **workspace**, which comprises the place where you are actually organizing and processing your work inputs. Within a workspace, you will find three functional components: your work area, your notebook, and your project manager. The non-digital equivalent to the work area is your physical desktop. This is where you would write, stack papers, sort articles, and perform your actual work. The next function is your notebook, where you would be organizing your thoughts and ideas pertinent to the work you are performing at the moment. The final function in this space is project management. Traditionally, this might be a whiteboard or corkboard on your office wall. It is here where you set your timeframes, goals, milestones, and other visual representations of your workflow.

The next category of the workspace is **storage**. In days past, this entailed both the file cabinet and a "looser" vessel such as a box, binder, or folder. The storage in the file cabinet principally maintains archival documents and reference materials that are vetted for relevance and future usefulness. Every document in the file cabinet got there because it was curated. The other, looser storage space acts as a temporary space for potentially relevant documents, clips, notes, etc. In the non-digital age, this box is where newspaper clippings, cocktail napkin notes, scratch notes, and other such items would go. Whatever goes into the container would eventually be looked at and evaluated for relevance to whatever work is at hand. Some of it gets organized into the file cabinet and some gets transcribed into a notebook. Much of the rest ends up in the trash. Where the traditional inbox acted as a space for correspondence, the loose storage box was an inbox for ideas.

Scheduling has always been the cornerstone of workflow management. This space comprises the *what and when* aspect of productivity. Hence, it has two functionalities: time management and task management. Time management is principally handled by a calendar – a tool with such ubiquity that the modern analogue is still a calendar. Task management can be something as simple as a list on a small sheet of paper. In tandem, these aspects are the basis upon which a productive workday is realized. Productivity is essentially knowing what you need to do and when you should be doing it.

The final category of the workspace is your assortment of **communication** channels. This is the space where technology has made the largest impact on the world of work and productivity. In the old days, communication was principally carried out via phone calls, terrestrial mail, or in-person discussion. Email and smartphones have transformed these first two avenues in obvious ways. Tech has even accelerated in-person discussions in ways unheard of mere decades ago. Text messages, chat clients, video calls have made such conversations as instant and inescapable as ever. More than ever, communications remain the principal source and accelerant of work – and must be treated as the largest input source in your productivity system.

We Are Already Cyborgs

Every component of the traditional workspace described above has a digital counterpart. In nearly every case, these implementations utilize web technology to become more robust and interconnected than previously imaginable. Combined with the power and ubiquity of smartphones, laptops, and other devices, we are living in a world science fiction writers conjectured only decades ago.

Futurists imagined what it would be like to have augmented brains. In truth, we are living in that reality already. Though not *quite* as impressive (or invasive) as cybernetic brain implants, the tech we use every day grant us many of the same superpowers. As a device nearly everyone has with them 24/7, smartphones are substantially augmentations to our minds. These devices can run spreadsheets, calculators, databases, communications, and access the entirety of human knowledge – all in the palm of our hands and on our wrists. Never in history have ordinary people had access to such amazing abilities. When leveraged correctly, technology can act as more than mere conveniences, but as full extensions to our brains. This is the promise of the technological future, made manifest today – no cybernetic implants needed!

In terms of work and productivity, modern technology offers unprecedented potential when leveraged effectively. For one thing, this tech can be utilized to make sense of the massive influx of information and communications we face every day. Technology gives us the superpower of being able to organize and process these inputs in an efficient manner. Access to external resources and knowledge necessary to complete tasks and solve problems is now but a few taps or keystrokes away. The ways in which devices and software interconnect allow for the utilization of multiple tools in tacking problems. It also opens the door to vast pools of aggregated data that we can leverage for effective analytics in our performance and productivity. Through data analysis, we can know more about ourselves and our habits than ever before.

Thanks to cloud technology, our ability to "remember" and retain information is incredible. In the past, we had only our organic brain, a pencil, and perhaps a bulky camera or tape recorder on hand to retain information. Research was compiled with paper clippings and bookmarks. Today, retaining documents, notes, ideas, brainstorms, recordings, news clips, articles, etc. can all be performed with a single click or swipe. Entire classes of repository software exist to store, sort, manage, and recall this information easily –independent of any specific device thanks to cloud storage. Used correctly, this easy and accessible tech effectively functions as extended memory for our minds.

Tech Utopia, Work Dystopia

Though we have the same brains as our ancestors, our everyday tech expands our brain power to process unprecedented amounts of information. These technological wonders enable extraordinary human connectivity and problem-solving. However, technology of any kind is rarely absolutely good (or bad), which hinges on how or if it is utilized. Regarding work and productivity, the problems tech solves are ones that were caused by technology in the first place!

The tumultuous times of the first quarter of the 21st Century lay bare much of the downside to this new technology. It was during this time, the internet matured from a burgeoning novelty into the cornerstone of daily life for billions (and created an underclass of people who do not have access). Online misinformation spreads faster than facts, which has fostered the re-emergence of extremism and authoritarianism. Digital communication depersonalized human interactions and created the toxic culture the internet is infamous for. Faster information and communication meant faster commerce and business – which resulted in more demanding deadlines and workloads for workers. The remote work revolution turned the home space into an extension of the office. In many parts of the world, wages did not rise with this increase in productivity. Perhaps the most apparent negative change is how in nearly every public or social space, many (or most) people are more engaged with their personal devices than each other or their surroundings. We live in an age of sci-fi wonders – complete with accompanying sci-fi nightmares.

In this new age, the key to work and personal productivity is to find the sweet spot in how technology fits into our lives. This tech avails us to resources and tools that were once regarded as fictional superpowers. The downsides to this technology ultimately stem from overuse and overreliance. To truly master productivity, you must develop the ability to see what is useful and what is wasteful in this space. True productivity is only realized through the *effective* use of technology – which often means relying on it as little as possible while maximizing its utility. This is a learned skill. The first step is to recognize both the immense power technology provides us while being ever aware of the potential dangers. Fear not – upon finishing this book, you will have gone a long way into developing this skill.

Technology will Save us from…Technology

Chapter 2: Essences of Productivity

Many people believe that all there is to productivity is to "just do it" with minimal planning – because planning is not doing and thus a waste of time. In days past, such a reductive approach would be adequate for most people to get things done. Today, the sheer volume and speed of information that hits us requires us to take planning seriously. "I am being pulled in so many different directions that I don't know where to begin" is an all-too-common sentiment for this very reason. The only way to effectively tackle the demands of work and life in todays' information-dense world is through proper planning and strategy.

In the purest sense, productivity is created through strategic use of your time. Strategic thinking is the opposite of reactive, unorganized, seat of your-pants thinking. It is a structured approach to work wherein every task and responsibility is addressed holistically. Rather than taking every new input and adding it to the proverbial work "pile", every input is carefully (and efficiently) evaluated and placed into an organized workflow. Many people make the mistake of being overly stringent in their planning. A critical trait of strategic productivity is flexibility - it must be able to adapt to sudden changes or emergencies.

As strategic approach to productivity poses and answers the four basic questions of effective time management:

- What do I have to do?
- When should I be doing it?
- How do I do it?
- Did I do a good job?

For any productivity system to be effective, it must address and answer these questions. Beyond this, a productivity system must be elegant, easy to utilize, and self-correcting. Unnecessary complexity will always doom a productivity system in the end – with potentially disastrous consequences.

Before going further, it bears noting that it can be just as easy to overthink productivity as it is to underthink it. Modern software tools and devices make it extraordinarily easy to foist new layers of complexity onto any system. Such complexity is often driven by novelty over convenience. It is not always easy to make the distinction. Being cool and flashy has a way of blinding even the most analytical among us. It is important to learn to look out for unnecessary complexity as you formulate your productivity system. This is also part of the self-correcting aspect of an effective system.

I find that an effective way of detecting and avoiding needless complexity is through hierarchy. When evaluating a piece of my system, I ask the following question: Is this piece part of an extant component of my system, or does it add an additional layer to my setup? If it does not fit into a current part of my hierarchy, there is a good chance it is adding complexity. If it is found to be adding complexity, the next step is to determine if it adds value to my system or not. This will become clearer as we progress further in this book.

Components of Productivity and Organization

A proper productivity system has three complimentary parts: Processes, Tools, and Behavior. All three parts are interdependent and essential to the functionality of the system. Recall that the purpose of a productivity system is to process inputs into outputs. The three components of the system work together to capture, organize, understand, and take action on inputs – converting them into desired work outputs.

The **Processes** component is the framework for how work inputs are processed. This component is further broken into two parts – procedures and routines. **Procedures** encompass the actual steps taken in performing your strategic process: Capturing, parsing, prioritization, planning, and action. Complimenting procedures are **Routines** – your scripted actions taken to ensure work inputs are properly and efficiently processed (via your procedures). This component entails capturing inputs when they come in, processing them, and regular purging of extraneous elements. In short, your processes are your predetermined steps for doing work while your routines dictate when and how these processes are performed.

Enabling your processes are your **tools** – the aspect of productivity most connected to technology. Recall earlier when we outlined the archetypal work tools and their general categories: Workspace, communications, storage, and scheduling. Every one of the essential work tools in these spaces have digital versions which enable automation and connectivity between them. Digital notebooks, task managers, project managers, file systems, and calendars comprise the backbone of modern productivity systems. It is the powerful data sharing integrations these tools provide that enable every aspect of your productivity system to operate as an extension of your mind.

The final part of the productivity trifecta is **behavior**. Despite the technological wonders of our time, we are still human and thus must contend with the biological challenges of productivity. We all deal with matters of energy levels, focus, attention span, and motivation in our pursuit of effective work. In the context of productivity, our behavior is comprised of our habits, attitude, your body, and your environment.

In summary, effective productivity is built upon the interplay of our processes, tools, and behavior. Our behavior dictates how effective we execute our processes, which are performed via the tools we put at our disposal. Only when these three components are optimized can we realize our most productive selves.

The Wide World of Productivity Systems

All work involves the transformation of inputs into outputs through action. Productivity systems are a means by which inputs are turned into actions. Life rarely avails us the ability to control the quantity and timing of work inputs – be they work tasks, personal responsibilities, or even hobbies. Fortunately, we have agency over how we process and manage the inputs that come our way. Technology is only increasing the volume of inputs in every aspect of our lives. A productivity system acts as the funnel and filter through which life inputs become manageable.

There are a scores of approaches and methodologies to productivity. Many emerged with the proliferation of the internet while others have their origins in the pre-digital age. Many of the concepts of these systems are non-exclusive and can effectively be mixed and matched based on personal preference. Even the productivity methods emphasized in this book can be supplemented or substituted with concepts from competing systems. Remember – a productivity system is intended to serve the user and not the other way around.

There are some systems of productivity that have proven so effective that it would be negligent to not mention them in a book such as this. Below are some (but certainly not all) of the most influential. Note that many tend to focus on one specific aspect of the inputs-to-action journey.

The Eisenhower Method

Possibly the most well-known example of task management is the Eisenhower Method, also referred to as the Eisenhower Matrix. This technique is actually a creation of author Stephen Covey based upon an anecdote by former president Dwight Eisenhower: "I have two kinds of problems, the urgent and the important. The urgent are not important, and the important are never urgent." Covey adapted this observation to develop a decision matrix for assessing tasks. In applying this matrix, a task is assessed for two dimensions – its urgency and its importance. These two dimensions result in four distinct categories (or quadrants) any task can have. By placing a given task into a quadrant, its respective priority is determined and thus how/when to take action in on it is decided.

Pomodoro Technique

Named after the Italian word of "tomato", this time management technique is based around splitting work time into small intervals. The typical execution of this has 25-minute periods of focused work followed by 5-minute breaks. A longer 15-minute break follows every 4 of these cycles. This productivity method has been in popular practices since the 1980s and has proven to be an effective way to maintain work focus for many people.

Deep Work

The creation of accomplished writer and professor Cal Newport, Deep Work is a popular approach to focus and prioritization. A key concept to this approach is the distinction between work that creates value vs incidental and clerical work. In Deep Work practice, learning to make this distinction is a critical step in determining where one's time and energy should go. Next, Deep Work fixates on methods to create and retain hyper-focus on these important work tasks. A vital part to this focus revolves around creating a distraction-free environment and minimal avenues for distractions to emerge. An experienced Deep

Work practitioner can leverage large chunks of their time and energy into intensely focused and productive work periods.

GTD – Getting Things Done

Few productivity systems are as popular as Getting Things Done, or GTD. This system is effective for many people, and it boasts a huge number of (very) enthusiastic devotees eager to profess its efficacy. GTD popularized the concept of capturing and processing inputs that has formed the backbone of nearly all productivity systems since its inception. This system also popularized the ever-handy "two-minute rule" for tasks: if it can be completed within two minutes, then do it immediately. For all other tasks, this approach focuses heavily on compartmentalization – particularly regarding context. This allows for practitioners to know what they should be working on at any given situation. In addition, GTD emphasizes task and information cataloging and recall. This makes it useful for preparation for future tasks and events. GTD was conceived just as the internet was finding its footing (after the dot com crash), and so was created principally for an analogue workflow. In recent years it has been updated to embrace more modern digital tech.

Inbox Zero

Overflowing emails is often the most conspicuous pain point for the overburdened knowledge worker. The term "inbox zero" perfectly describes the approach it takes with email management. Inbox Zero is not so much about keeping a completely empty inbox as it is optimizing its use. Emails are to be checked only once or twice per day. It is focused on optimizing time spent on emails by aggressively processing them out of the inbox. This processing borrows heavily from GTD: any message that can be resolved in two minutes or less are addressed immediately. Everything else is either delegated, deferred, or deleted. The great thing about Inbox Zero is that these concepts can be applied to inboxes of all kinds – not just emails.

Essences of Productivity

Assessing productivity systems can be as daunting as the task backlogs they are intended to mitigate. The approach to productivity that is the focus of this book is yet another one of these. This particular system is intended to be broad and covers the entirety of the considerations and requisite skills of modern productivity. In this way, even should you choose a different route in adopting your own productivity system, reading this book will make you well-versed and aware of the intricacies of the topic. As the adage goes, take what is useful to you and discard the rest. Good luck and let us go forth.

All Success Begins with Planning

PART II
Tools

Chapter 3: Understanding and Utilizing Tools

The biggest difference between modern productivity and the old days is the assortment technological tools at our disposal. It is this tech that augments our ability to collect, process, organize, and act at levels far beyond what our brains can do on their own. A properly utilized array of technology is what grants us the cyborg superpowers of science fiction. However, a poorly implemented jumble of tools can do just the opposite - making you as productive as a rock.

Technological tools should not dictate the fundamentals of your productivity system. However, as the implements by which you realize your system, they do define its scope and limits. They will also inform many of the subtleties of your system – such as how certain information and tasks are passed from one stage of your process to the next.

Before discussing the processes of productivity, it is important to understand the categories (or roles) of technological tools that comprise the working elements of the system. Nearly every productivity system worth its salt will utilize these types of tools. Each of these roles have numerous solutions to choose from – with more available every year as the burgeoning productivity industry expands. In many cases, one service or implementation can fill multiple roles. This is fortunate as it is preferable to keep your assortment of technological tools as small as possible. The more elegant a system, the more effective and sustainable it tends to be. Interconnectivity is also fortunately becoming increasingly standard with these tools. There is also a thriving cottage industry of services that exist to help bridge these platforms together. This means that it is possible to have an effective productivity system that utilizes a broad array of tech platforms – just so long as you are up to the task of managing these connections properly (this is probably best left to experts and not beginners).

Know Your Roles

Recall that all the basic components of an effective work setup originated with non-digital tools that were replaced by (usually) superior digital manifestations. You will see here that every tech solution you integrate into your productivity system will fill one or more of these. Productivity systems must utilize tools for handling communications, workspace, storage, and scheduling. In every case, the technological implementations of these will be served by software – be it an app, program, and/or service. All software must run on a device of some kind, and so your productivity toolset must include one or more devices that best serve your needs.

Utilizing Devices

The combination of persistent internet connections and portability is what enables modern tech to act as true augmentations to our brains. This portability and power allow us to access these services and apps virtually anytime, anywhere. Today, we can easily capture media, take notes, set calendar events, and set reminders anywhere we go. This is to say nothing about our ability to communicate with people anywhere on the planet at any time, or the fact that we can access the entirety of human knowledge from the grocery store. We take this power for granted today when it was fantasy only a few decades ago.

As with any tool, inefficiencies can emerge when they are poorly utilized. As powerful as the internet is, it is also the biggest source of distraction and wasted time in the history of civilization. Enough hours were already wasted looking at cat pictures over a phone modem on the bulky family desktop. Now, entire days of productivity can be lost as we can look at internet memes on our phones from practically anywhere. And this says nothing about the advancement of phone videogames, which are now capable of replicating the fidelity of console games. The smartphone is simultaneously the greatest advancement and hindrance to personal productivity there is.

You simply cannot design an effective digital productivity system without careful consideration of your tech hardware. Gadgets are fun and exciting - especially when they are new. However, this mentality often traps people into having an overdependence and overabundance of tech in their lives. This inevitably leads to wasted and inefficient time usage. In considering your tech needs, never forget that technology is supposed work *for* you by making life easier and more efficient. Your tech should create minimal distractions or extraneous complexity in your life. With every piece of tech in your life, ask yourself if it brings productive value into your life or not.

Optimizing the role of tech in your life should be focused on making its usage as effortless as possible. Any input, idea, note, communication, etc. should be able to be captured into your digital system with ease. Curation and recall of this information should also be clear and intuitive. Your devices are what you use to access the power of your digital tools. Inadequate hardware can diminish or even eliminate the benefits these tools provide.

A common pitfall faced with tech is having too much or too little of it. It is most common to have too much – especially for those of us who are enthusiastic about the novelty of shiny new devices. This is an easy rabbit hole in which to fall, given how many cool gadgets available these days. I personally fell deep into this trap at one point. My digital setup included a desktop PC, two tablets, two laptops, two smart watches (don't ask), and multiple voice assistants. Aside from being expensive, this setup added many unnecessary layers of complexity and inelegance to my system. When I had to capture something, I had to take the extra step in deciding which device would be used to do so. In processing my inputs or doing work, I had the extra complication of ensuring every device was synchronized. Few things were as disruptive to my workflow as having to figure which of my computers was the most up-to-date ready to work. Worst of all, my workflow had become dependent on using all these devices in a given day and so if one was unavailable, I had to waste more time by improvising. Take it from me – the drawbacks to having many devices defeats nearly every benefit. As with so many things in life: keep it simple.

Ultimately, in an optimized digital productivity system, all you really need is a device for capture and another for processing and working. A smartphone and a (decent) laptop are the most basic and elegant device setup and can serve most people quite well. Obviously, some types of knowledge work require a powerful desktop or a touch screen. However, the basic smartphone and laptop combo demonstrate the fundamental capabilities hardware must fulfill in a digital productivity setup: convenience, power, and portability. Your smartphone, by virtue of its size, must serve you by enabling quick, convenient capture of any inputs you encounter when away from "core" computing devices. As powerful as phones are, core computing device like laptops and desktops offer the extra power and superior input devices (keyboard, mouse, etc.) to enable effective parsing of inputs and execution of work. Finally, portability is key in that it frees you from being tied to a single location – not only freeing you from limits to where/when you can do work, but it inoculates against situations where you might be unable to access your office or workplace.

Everyone is different, and such a spartan hardware setup might not be the best for you. The key takeaway is to be mindful of what you need hardware-wise and always strive to be as reliant on as few devices as possible. In my case, I reduced my overabundance of tech to just four principal devices: one smartphone, one laptop, one tablet, and one desktop computer. My work often involves spreadsheets and graphic design, so I utilize a powerful desktop to accommodate. My laptop is less powerful but grants me the flexibility to work off-site when I need to. My work also involves a great deal of reading, and my preference for a tablet for this justifies maintaining a separate device. Again, your mileage might vary – just remember the "less is more" mantra applies to every aspect of a digital productivity system, and hardware is no different.

Leveraging Integration and Automation

An effective digital productivity system is one that requires as little effort as possible to utilize. As a matter of synergy, the tools that compromise your system must be configured to meet this requirement. To achieve this, your grouping of tools must meet three objectives. First, the functionality of each tool must be intuitive and elegant. You would already have accounted for this in your assessment and selection of your productivity tools. The next objective is to make use of automation as much as possible. Many productivity applications and services have built-in features that can automate some or all parts of your processes. Ensure that you look for this functionality in your selection process. Finally, your tools should be interconnected in such a way that data, tasks, and other information can be shared between them as easily as possible. Increasingly, many applications in this space offer built-in connectivity to other services either directly or via an API. Automation and interconnectivity enhance the latent usability of your tools and serve their ultimate purpose in your system: minimal effort for maximum utility.

The continuing proliferation of digital productivity applications has given rise to additional services that enable extensive automation and interconnectivity between many platforms. Services in this space include IFTTT, Zapier, Pleexy, Automate.io, and Power Automate. These services allow you to connect multiple applications and services together and share data between them. Automated processes can be created using conditional statements in the vein of simple computer program logic. For instance, you can create a process where if you flag an email in your email client, it will trigger a process that automatically creates a corresponding to-do in your task manager. Another example: when your name is mentioned in a Slack conversation, it triggers a process that creates a log of the mention in a Google Document. These are very basic examples, as the power and flexibility of these services expand regularly. Some of the more robust automation & connection services include extensive filtering and advanced actions (like updating profiles, sending emails, pushing notifications, etc.).

To utilize an external automation & connectivity service is generally the mark of a more advanced productivity setup and is not for everyone. These services are powerful but are not always the easiest to implement efficiently. Also, as with all "extra" tools in a digital system, they add extra layers of complexity in how they are setup. The power of these services can greatly simplify busy workflows, but only if their implementation does not exceed the amount of work they reduce. Bear in mind that these services are, like all software, imperfect and can sometimes fail or misfire.

The more software and services you are dependent on, the more potential points of failure you are creating. Ultimately, these services are best left to power-users or those with unusually large abundance of inputs flooding their workload.

These automation & connectivity services are essential in cases where you need to enable your digital tools to connect to each other. Ideally, you want to utilize built-in connectivity for your productivity tools. However, in cases where you have a tool or service that does not have such functionality, then it justifies utilizing an external automation service. The need to be able to share information between your essential tools is just too great – your email, task manager, calendar, repository, notebook/whiteboard, project manager, and file systems MUST be connected. Cutting & pasting or manual transcription is simply out-of-the-question and completely antithetical to an integrated and effective digital life.

Use Only the Right Tool for the Job

Chapter 4: Communications Channels

Undoubtedly, one of the largest sources of work inputs for most people comes through communication channels - be they assignments from bosses, collaboration with colleagues, or requests from clients. For many of us, communications are veritable floodgates of things that require our attention. They also tend to be rife with irrelevant, unimportant, or redundant chatter that makes it difficult to pinpoint the truly important stuff. Of all the technology that comprises modern workspaces, communication is where you likely have the least amount of flexibility on which you can utilize. Businesses and organizations often to dictate which communications platforms they (and you) utilize.

Email

Email continues to a principal method of communication for most organizations. If you are fortunate, you have carte blanche to select the email client that is best for you. Even if you do not, most email clients have feature sets that allow you to effectively parse and organize your messages. In considering an email client, your main concern should be on how well you can organize messages – ideally in as automated fashion as possible. The essence of managing large amounts of email (or any bulk of information) revolves around metatextual identifiers –more colloquially known as tagging or labeling. Though different email clients might use different terminology or methods of implementation, the result is the same – emails can have descriptive labels attached to them that allow them to be grouped and sorted.

Like all bulk information handlers, emails are best managed by utilizing tags, folders, filters, and search. Every major email client has incarnations of these features. Though not always 100% identical in implementation, the nuances between providers primarily lies in naming differences. Tags, or labels, are just that – descriptive marks that describe general properties of the message. For instance, an email can be tagged by client name, work type (phone call, clerical work, etc.), importance, urgency, etc. Tags differ from folders in that an email message can have multiple tags applied to it rather than a single folder. Folders are groupings of messages with the same general context or category – an email can have many tags but only one folder location.

Over the years, email clients have become more focused around search and filtering as email accounts grew in size and complexity. You will need to have a means of quickly locating emails with certain combinations of tags and other properties such as subject matter, sender(s), or keywords. The best email clients have robust search capabilities, though they may implement search in slightly different ways. In choosing a client, be sure to evaluate how its search functions work and account for which ones are more intuitive for you. Filters are essentially searches that you can save as shortcuts in the client. Filter functions are where the power of effective tagging can be seen. By combining diligent tagging practice with effective search & filtering, you will be able to process and manage large numbers of emails with ease. As with search, be sure to take a close look at the filter functionality of any prospective email client to see which is the best match for your needs.

Collaboration

The onset of the COVID-19 pandemic accelerated an already active adoption of non-email communications channels across organizations of all sizes. Chat clients were once thought as the solution to the communication clutter email had created. In some ways, it worked in that it combined the best of both worlds of direct and asynchronous chat. The pandemic also pushed video conferencing into nearly every home office overnight. Chat and video clients attained such a quick ubiquity in knowledge work that every suite of work software now has this functionality.

Ironically, the proliferation of video and text communication channels added more to the clutter than it resolved. Email never went away, nor was its importance diminished. Furthermore, workers find themselves with few choices on whether they engage with these channels – let alone the option to select which of these platforms to utilize. In terms of managing a productivity system, it falls primarily to using whatever tools these platforms offer.

Effective management of non-email communication channels comes down to your routines, internal tools, and external tools. As we will cover later in this book, your productivity system will utilize multiple methods of capturing relevant information. In most cases, this information translates into three general types of inputs: events, notes, or tasks. As you parse through these communication channels, you will know how to identify these inputs and have a routine in place to quickly capture them and place them into their right place for further processing. As you should be taking regular time to engage with these channels, your routines for capturing relevant inputs should be a regular habit. Many of us are inundated with so much chatter that it might be prudent to consider automations. Chat and collaboration suites have built-in tools for implementing this. For instance, Slack – one of the most popular of these types of services – has a suite of automation tools that can help identify and sort relevant information from bulk conversations. In addition, there are numerous external services that connect to these clients that provide additional means of parsing through messages automatically. Again, organizations sometimes block access to these features, so check with administrators before planning to integrate them into your workflow.

Be Available but Not Overwhelmed

Chapter 5: Your Auxiliary Memory

The advancement of internet technology into our work and personal lives brought with it an unprecedented volume of information for us to contend with. In addition, the process of knowledge work entails the creation and management of countless computer files and other digital assets. Modern-day horror stories like important files being lost or screens filled with hundreds of icons are part of the zeitgeist. Managing digital files and information has become an essential life skill and marks the difference between those who thrive or struggle in the digital age.

Effective digital storage encompasses two main considerations: file organization and information recall. Organizing your files means just that – ensuring that your digital assets (documents, pictures, and everything in-between) are safe and easy to locate. Information recall entails the skills and methods you use to capture, organize, and retrieve data with as little effort as possible. Though both practices involve the storage of digital assets and information, they serve different usage cases and thus must be handled by different types of tools and solutions.

The Information Repository

Reductively speaking, life is comprised of streams of inputs that demand our attention. Ideas, tasks, events, reminders, brainstorms, and news comprise the information we are subjected to all day long. These inputs are highly variable in terms of their content, context, importance, urgency, and value. One of the primary purposes of a productivity system is to break these down into manageable and organized pieces.

Before they can be parsed and turned into actions, inputs must be captured to one of three places – your calendar, task manager, or repository. This is because at the highest level, an input will either be a **task**, an **event**, or a **note**. Events are unambiguous and can be captured directly to your calendar. Tasks are inputs that are overtly actionable and should be captured to your task manager. Everything else goes into perhaps the most crucial part of your digital toolset – the Repository. The **Repository** tool acts as the central location where you capture, organize, and store work inputs. It is the singular unified inbox for all the information you encounter every day. I feel that the term "repository" is the most apt for this role in that it defines it role as a place for bulk storage and processing. In other productivity media, you might see this role referred to as "the inbox" or "notebook".

There are many competing solutions and services that can occupy this vital role. Digital notetaking services fill this role nicely, as many of them are built around ease of capturing notes and media easily. The increasing assortment of hybrid productivity apps are also making inroads into this space. These solutions find their efficacy in that they honor the three most critical characteristics of what a proper repository tool must have: Easy capture; Effective organization; and Effective action.

First and foremost, the repository tool must facilitate easy capture of inputs. It should be equipped to handle inputs from a wide variety of sources, formats, and types. Inputs can manifest as text, handwriting, audio recordings, photos, videos, screenshots, website content, and emails. Your tool of choice must be able to easily accommodate all these formats in a unified location. The best notetaking services have apps, plugins, and other implements that can do this. In addition, many have APIs (basically a way for one app to share data with another) that can connect to virtually any device, program, or web browser to further expand this functionality. In selecting your repository solution, your goal is to empower yourself to capture every input you encounter at any point in your day with as little effort on your part. Given that nearly all of us possess smartphones, a repository tool must have an app that can be called upon at any time to capture any incidental ideas or other inputs you encounter daily.

The second critical aspect of an effective repository is its capacity for organization. All inputs should filter into an inbox area in your repository. On regular intervals, you will set aside time to process your captures to make sense of them and where they fit into your life and workflow. For instance, a web clipping you captured from your browser has specific relevance to a project you are working on. This clip should have a category or folder in your repository to which it can be placed– no captures are to remain in your inbox as they all will have a predefined location to "live" under. In addition, you might want to label this clipping with a tag that indicates its subject matter. The repository tool should also make it easy for you to make title edits and to highlight certain elements of the clip for easy recall later. An effective repository solution must accommodate easy editing, highlighting, tagging, and categorization. Finally, these organizational features must all be in service of searchability. A repository without a robust search functionality is practically useless.

Finally, a repository solution must empower your ability to act on your captured data. Action can take many forms depending on needs or context. As with your regular practice of processing the inbox within your repository, you will be making routine reviews of your organized inputs. For instance, as you are working on a certain project, you will be referring back to the assorted captures you made regarding that project or subject matter. For instance, a web clipping you captured last week may have particular relevance to what you are working on this week, and you want to review it tomorrow. A good repository tool will have ways for you to flag this clipping and even create a task in your task manager reminding you to do just that. In some cases, you might need to share this clipping with colleagues or copy it to your workspace directly. Again, your repository should have the ability to facilitate this with minimal effort on your part. The tool must serve your needs with the least amount of hassle.

Fortunately, there are many great services that can achieve all three fundamental needs of an effective repository. Among the oldest and most established is Evernote, which practically created the modern digital notetaking industry. Another stalwart in this space is OneNote, though despite its strengths is not particularly strong on the tagging side of things. More recent comers like Notion and Obsidian lean heavily into the bulk data curation paradigm – they focus heavily on tagging, metadata, and linking as would be done in a database. Remember the key functions a repository tool must address in any evaluation you make regarding these services.

Reading / Media Lists

The modern digital world provides a practically endless array of media of relevance to both our professional and personal interests. Part of thriving in this world is to be able to manage this assortment of content. It is counterproductive to consume all such media as it comes to us, and so it best to maintain a reading/media list for later use. It is more than adequate for this to be handled in your repository – simply capture every piece of media you want to circle back to and apply a tag (such as "ToRead" or "ToWatch") to create your list.

Much of the media of this type is disposable, and less likely to be important for you to reference later. If you are concerned with creating potentially undue clutter in your repository, or are a productivity power-user, there are other options just for this type of incidental media. There are numerous services that accommodate this very usage case: A place to store digital articles, audio, or video that you want to quickly save for later. They have many of the same capturing hooks that a notetaking app has, but the focus of the service is centered on curating a read/watch/listen later list. Paid services such as Pocket or Instapaper offer this type of content management with features to make consumption of captured media easy. In addition, many web browsers support read-later lists that are as simple as a right-click. One additional intermediary approach to this is to capture this type of media to your task manager and build your read/watch later list there. As we will discuss later, a good task manager will enable this type of functionality.

If you choose to use a supplemental tool such as this, beware that doing so means adding a layer of complexity to your digital life. Extra complexity is not inherently bad nor counterproductive, but only if it is properly implemented and brings actual value to your processes. To this end, maintaining a separate read/watch/listen media list should serve two purposes: The first is to make it easy to save content for later when you make specific time to consume it. The second is to reduce clutter in your repository – this is especially desirable if you consume large amounts of "disposable" media that holds little value for post-recollection.

As with any optional ancillary service or addition to your productivity system and/or toolset, interconnectivity is important. For example, as you review your media list you notice an article that turns out to be especially important to your current work project. If you are using an external reading list service, it is not captured in your repository. Fortunately, you have setup connectivity between this service and your repository, so the article can be sent over and captured there with a tap of a button. Or, if reading this article turns out to be a priority for you, you can tap a button that sends the article to your Task Manager software. The capacity to perform both actions are must-haves for any external service you utilize in your system.

We are Already Cyborgs

Chapter 6: Managing Files

Since the early days of personal computing, proper file management marked a divide between novice and expert. Initially, this entailed little more than managing files on a single hard drive. The cloud revolution has made it so that our files can effortlessly exist across multiple pieces of hardware and platforms. This helped reduce the risk of file loss – be it from user error or hardware failure. However, it also made proper file organization more important than ever: different devices offer different user experiences, and so your files will need to be curated in a manner that serves them all.

Even if you create comparatively few digital files and assets in your work, chances are you utilize a smartphone as your primary photo and video camera. Most device providers and cloud storage solutions can be configured to automatically backup photos and videos. This is an important feature that everybody should be using regardless of their other file needs – it keeps these important assets from being trapped on your devices. Devices can be easily lost and destroyed. Hedge your bets and get your photos and videos backed up ASAP.

The interconnectivity of digital work requires your file management system utilize cloud storage. It is the key to a fully integrated digital life that lives across all devices and workspaces. In creating your optimal file storage setup, you must consider the cloud service(s) you wish to use. There are many competing options in this space that can be grouped into three basic types. First are the platform-agnostic file storage solutions such as Dropbox. Such a service can be utilized on nearly every major hardware platform like Windows, Mac, iOS, Android, etc. and generally acts as traditional file storage. The next type are more hardware-oriented offerings like OneDrive or iCloud. Services like these are built to integrate closely with a specific operating system or hardware, with their full value and utility only realized when used on their preferred setups. OneDrive works best with Windows while iCloud is intended for Apple products. Though these services might work across other platforms, their unique value is relegated to their intended environment. The final type of cloud storage are the "built-in" services that accompany an app or service. Adobe, which provides prominent design software like Photoshop and Acrobat, offers integrated cloud storage with their products. Similarly, Google Drive is a cloud storage solution built into the G-Suite (or Google Workspace) assortment of services.

Which of these three types of cloud storage services is best? I recommend avoiding the use of app-provided storage.

These might be nice perks to said apps but tend to exist mostly to support the app it is packaged with. For instance, though the Adobe cloud storage works great with photoshop, you do not want to split the design files of a project with the other files living on your general storage. In addition, should you cease to use the app/service, you run the risk of losing these files. You want your cloud storage to be as centralized as possible and unhindered by hardware or OS constraints. Keep things as future-proof as you can. To this end, I recommend going with a platform-agnostic service like Dropbox. If you are particularly committed to a specific platform or OS, then go with what optimizes your experience – just be aware of any limitations that come along regarding competing platforms. For example: OneDrive plays well enough on Mac but iCloud is far more limited on Windows.

As wonderous as the cloud revolution has been, it introduced a slew of security and privacy concerns. There are no two ways about it – if your files are stored online then they are at risk of being hacked, intercepted, or otherwise accessed by others. All cloud providers will have some level of access to your files, which are spelled out in their terms of service. In most cases, they "respect" your privacy but will comply with all laws to which they are subject and might access your information as part of providing the service. This is standard across all such services. Cloud providers might offer different levels of encryption on files and are generally adequate for most users of the "big" cloud storage services. In recent years, several services have added special ultra-secure file directories for particularly sensitive files. As what should be a standard practice for the entirety of your digital life, these services also support two-factor authentication for logins.

Active Storage vs Archival Storage

Some of us regularly produce large volumes of digital files and data that creates a need for an extra level of organization. A well-managed digital file system of this type should have a distinction between active and archival storage. What is the difference? Active storage denotes the files and structure surrounding files you are either actively using, frequently accessed, or was created recently. Archival storage is for your files that are essentially the opposite – not recent nor likely to be accessed soon or on a regular basis. The principal reason to maintain a separate storage space for these files is to reduce clutter in your primary file storage. It also can enhance searchability and organization of your files in that archived files – a well-organized archive might benefit a different folder/directory structure than your active directories.

The point of a file archive is to separate the inactive files from the active ones, which can be achieved in several ways. The most straightforward way is to have a distinct directory to which these files are moved. This would be done on the same drive or cloud directory as your active files. If part of your archive strategy is to free up computer/device storage space, most cloud services allow you to set files or directories to be "online-only". These files thus only exist on your device storage while opened or downloaded specifically. The final way to manage an archive is to have it exist on a different location as your active files. This can be achieved by having your archive stored on a separate cloud storage account or service. If having your archive on the cloud is not necessary, you can also maintain it on external hard drives, though it is recommended to have at least one extra copy for sake of redundancy.

The need for a curated file archive is generally the concern for those of us with more expansive digital file outputs. For the most part, if your file volume is modest, then all you need is a single digital file directory. How do you know if you should have a separate archive? It generally comes down to three considerations. The first has to do with the volume and age of the files you have. If you have large numbers of files that go back many years, then an archive might help to reduce clutter. The second consideration has to do with available space. Storage services and hardware have technical and service limits on storage. Syncing large amounts of files can often take time, bandwidth, and hardware resources – moving archival files to a separate storage place mitigates this. The final consideration revolves around personal preference and organizational style. For instance, I prefer minimal clutter or complexity in my file system- and so I only keep my active and most recent files in my general file storage location. Everything else is archived to a separate location. As a redundancy, I keep my archive on a different cloud account than my main files.

The last aspect of digital file management concerns physical file backups. Though cloud storage continues to be reliable and relatively secure, never doubt the assurance of having a backup on physical hardware. Our lives become more and more defined by our digital assets every year. Cloud services are only as good as their uptime and accessibility. A regular practice of making backups of your files to a securely stored portable drive is an easy way to ensure peace-of-mind and integrity of your digital life.

There is No Such Thing as Controlled Chaos

Chapter 7: Time & Tasks

What is productivity without time and task management? It should come as no surprise that the associated digital tools for handling time and tasks are vital components of a productivity system. Remember that all inputs are ultimately either a note, a task, or an event. However, just because an input might be initially categorized as a note does not mean it cannot become a task or event at some point in the future. For instance, an idea you capture as a text note in your repository can become a task if, upon later review, it ends up being an idea worth taking action to bring into fruition. As every task ultimately must be acted upon, you will eventually be scheduling the time (an event) to do so.

Calendar

Of all the traditional tools of the workspace, the calendar has changed the least as it transitioned into the digital world. It is the classic tool of time management that has only gotten better as a digital tool. In terms of selecting the right calendar for your work style and preferences, nearly every option available has comparable features and competencies.

In a modern productivity system, the calendar fills a role far beyond a mere place to schedule meetings and events. It is the principal tool by which you will plan out your day and how your time will be delegated. As we will cover later in the book, time blocking is the primary way in which you plan your days and set boundaries between daily distractions and the meaningful work and effort you should be focusing on.

The digital age brought several important innovations to the calendar. Perhaps the biggest improvement was the ease in which events and schedules can be created and altered. Digital calendars have hooks that connect them to email platforms as well as a variety of services – from task managers to CRM (customer relationship management) software and everything in between. These hooks make it easy to automate schedules and minimize the chances for important events and tasks to slip through the cracks.

Though feature sets are so similar across various digital calendars, there is one caveat to consider when selecting one: Some providers are less generous about which platforms and hardware their service supports. Apple, as it does with much of its connected productivity apps, is far less platform agnostic than offerings by Google or Microsoft. Again, it is advisable to go with options that have as few hindrances as possible. Your mileage may vary.

Task Manager

If productivity is about getting things done, then it begins with knowing what those things are in the first place – hence the task manager. The linchpin of all productivity tools is the **task manager**, the modern descendant of the humble to-do list. The power of technology has turned what was little more than lists on a notepad into some of the most powerful productivity tools available.

The assortment of task management solutions is quite large. This industry continues to grow as newcomers emerge while existing productivity suites continue to add task functionality. To complicate matters, task managers tend to have significant differences between them. There are many different theories on how to best organize, structure, plan, and execute tasks. Most task managers specialize in one or two of these approaches the ones that try to accommodate all task management styles usually end up failing to excel at any of them. Some of these task structures include: **Kanban**, a visual approach where lists are sorted side-by-side usually grouped by timeframe; The aforementioned **GTD**, which groups tasks based on contexts; **Work Breakdown Structure**, which break tasks into progressively smaller and specific subtasks. Notice that these, and many more, approaches to task management are closely tied or derived from various project management methodologies.

There is no universal "correct" way to manage tasks, and each approach to task management is suited to different work styles and preferences. The first step in choosing which is the right solution for you is figuring which theory of task management is your best fit. Hopefully, you have sufficient freedom in your work to make this assessment. Some organizations foist a specific type of workflow on their entire workforce. Even in such cases, there are ways to maintain a *personal* productivity system that fits your style even if you are compelled to work under a system that is less so. This will be discussed in detail further in the book.

As with every other item in your productivity toolbox, your task manager must accommodate your work style and support connectivity to your other tools. The leading solutions in this space include: **Todoist**, which offers powerful filters and task hierarchy; **Microsoft To-Do**, a task manager integrated into their Office suite; **Google Tasks**, the G-Suite (Google Workspace) task manager; **TickTick**, which specializes in integrated calendars and additional habit tracking features; **Trello**, a Kanban-specific task manager that has close integration with the Jira suite of project management tools. Aside from these dedicated task managers, many project management and digital notebooks have integrated task management functionality. Services like Evernote, Notion, and Basecamp all contain this functionality at various levels of sophistication. Most serious task managers offer API hooks and connections to other tools and services – a must-have feature for any integrated productivity system.

Time Cannot Be Created, But It Can Be Managed

Chapter 8: The Workspace

. Think of your digital workspace as you would a chefs' kitchen. A good chef has all their ingredients, appliances, and tools organized in pantries and drawers. When they are cooking, their countertop is where they arrange the necessary tools and ingredients and process them into finished dishes. This work area is dynamic and exists to serve the needs of the chef for the food they are specifically preparing. When they have completed their culinary projects, the kitchen is cleaned, and everything is put back where it belongs – all to be ready and organized for preparing the next meal. Regard your digital workspace in this same manner.

Project Manager

In many ways, personal productivity is personal project management. It helps to think of your productivity system as if you are project managing your entire life. Your hobbies, home maintenance, pet care, finances, and even relationships can be successfully managed as you would a traditional work project. Hence, one of the invaluable tools of a personal productivity system is a project manager of some variety.

Many people conflate project management with task management, which are not the same thing. It is important to accurately understand this distinction because task managers can be used for project management and vice versa – making it easy to conflate the two. Project management is a process in which a specific outcome is planned out and broken into a series of steps in a hierarchical manner. Projects have clearly defined start, middle, and end points where the outcome is unambiguous and measurable. Out of every project comes a list of tasks to be performed in service of said project. This is where task management comes in. Task management is more holistic in that it involves the management of tasks from multiple projects as well as any number of incidental tasks big and small. Project management determines what tasks are needed to achieve its goals. Task management organizes these tasks in a way that they can be effectively executed.

In terms of features, many task and project management tools are capable of filling both roles. However, the gulf between the best task managers and best project managers is significant – there are few solutions that excel at both. This does not mean that they both cannot be served by the same tool. As always, your mileage will vary depending on your needs and tolerances.

Most project management solutions are primarily aimed at teams and more enterprise-level needs. Many still offer single-user tiers, including free accounts, but tend to lock features and functions behind paywalls. In addition, the hybridization of productivity software and services means you have more choices, but requires more research to find the best fit. This is also another space where you might be tied to an organization that is tied to a specific project management suite – eliminating your choices outright. It bears noting that, especially for personal project management, specialized tools are not necessarily. If you know enough about the fundamentals of project management, you can get by with bulleted lists in a text or spreadsheet program. Some of the players in the project management space include **Jira**, **Notion**, **Microsoft Teams**, and **Basecamp**. Increasingly, the feature sets of task managers and notetaking apps are integrating project management functionality. This trend has been going on for some time and shows few signs of slowing down.

Digital Notebooks and Whiteboards

In the context of productivity, the workspace is the central hub where all the other elements of the system meet. Notes, ideas, and media are captured and sorted neatly in your repository. Your task manager lists everything you intend to accomplish. The time has been blocked out in your calendar to work on the specific project at hand. Ideally, your email is closed, and your phone is muted. It is now time to actually do work.

Work of all types needs an area for it to happen in – a place to arrange ideas and materials relevant to the tasks at hand. Aside from the software required to create work outputs (documents, spreadsheets, compilers, etc.), it is important to maintain a space to sort, plan, and log knowledge work as it happens. In the digital realm, this role is principally filled by digital notebooks and whiteboards. While still largely a matter of work style, utilizing a connected online note space in this way marks one of the principal ways in which modern tech has the capacity to supercharge work efficiency and efficacy.

This aspect of the workspace comprises both the functionality of a digital notebook and a digital whiteboard. As a **whiteboard**, it is where concepts and components of your work can be visualized and organized. Experts from fields as varied as software development, writing, or graphic design make use of the power of whiteboards every day. As a **notebook**, this component of digital productivity is where you record your ideas and coalesce research surrounding your work tasks. Both usage types center around arranging thoughts and concepts in both text and audiovisual forms – the principal difference being that whiteboards focus more on abstraction and visualization and notebooks on long-form inputs. The differences can be vague, depending on the specifics of the work at hand. Fortunately, and for this reason, nearly every digital notebook worth its salt functions just fine as a digital whiteboard. In fact, as notebooks become more powerful and feature-laden, the differences become increasingly fuzzy. The best thing about these tools is that they support multiple types of media including text, images, tables, embedded documents, and multimedia. That these services sync across multiple devices via the cloud is a cherry on top.

It bears noting that many of the same solutions that fill the role of a digital whiteboard/notebook are the same as those that fill the digital repository role. Again, this is a place where one tool can fill multiple roles quite well. However, as the roles between repository and workspace are quite distinct, care must be taken to separate them. The repository role is best thought of as a curated collection of research notes, clippings, scratch, and multimedia materials. The workspace is where the work surrounding these materials is condensed and abstracted as you conduct your specific work objectives. To illustrate this with a pre-digital analogy: The repository is the file-cabinet of clippings and notes you have sitting by your desk as you arrange your work on your office whiteboard and personal notepad.

Regarding digital notebooks, how you make the separation between your information repository and workspace comes down to preference. Personally, I keep my repository and workspace on separate platforms. I use **Evernote** as the place where I capture, arrange, and maintain my bulk digital repository.

Evernote is my file cabinet. For my workspace, I utilize **OneNote** due to its superior capabilities as a digital whiteboard. As with every other tool in my digital arsenal, I ensure that there are always ways to quickly find and attach or export relevant content from my repository to my whiteboard. This arrangement works incredibly well for me. If you do not have the stomach for this extra complexity or do not capture a huge bulk of information (I capture hundreds weekly), keeping things on one digital notebook will work just fine.

To Do It All, You Must See It All

PART III: Processes

Chapter 9: Productive Living in Action

In the learning processes, it is helpful to see an example of the end-result as a reference before exploring their inner workings. Accordingly, I will be describing how my personal productivity systems function in a typical day. Many books on this subject seem to exist as vehicles for the self-aggrandizement of the author and their lifestyles of achievement. I even found a productivity book where author boasts of how his system assists his sexual conquests (cringe). With humility, I present my personal systems strictly as an example of what I have found works best for my needs and temperament. It is the result of much research and experimentation and so I stand by its efficacy as a well-managed digital life. Undoubtedly, as you continue with your own research and experiments, your optimal systems will differ based on what works for you. For now, please accept my anecdotes and expertise as simply a guide in your personal journey in this space.

The End is the Beginning

For the most part, the quality and productiveness of my days are determined by my routines the night before. It is at the end of the day where my next day begins. This daily review is part of nearly every productivity system out there for a reason – it works. My daily review is where I analyze and process the days inputs, including notes, tasks, emails, and everything else on my plate. It is the last time of the day that I think about any work-related matters or personal responsibilities. This crucial step to my day immediately precedes the final stage of my day, which comprises my daily workout and the time I save for "unproductive leisure" (aka video games). As a matter of maintaining the habit of this routine, I will not start my workout or leisure activities until the daily processing is finished.

The first step in this process is to review all my communications channels. For me, this is primarily in the form of emails. My intention is to have as few emails in my inbox as possible and to make sure nothing has fallen through the cracks. Utilizing the star and label functions in Gmail, every email that comes to me will be one of five types: messages to delete, archive, reply, action, or retain. Utilizing the Zapier automation service, any email that I flag for reply or action will automatically have a to-do created in Todoist, my task manager. I lightly use Slack and Discord as alternative communication channels and use a similar approach in how I process the days messages.

Once my communications channels are clear, my attention goes to my notes and captures. My work involves significant research, which results in frequent clippings of articles and other media. These are captured into Evernote, my preferred repository tool – my digital file cabinet. In addition, I use Evernote as the place where my incidental ideas and insights are captured. It is also where I place receipts, manuals, and other such documents that I want to exist in a portable, searchable place. Every item I capture to Evernote goes into an inbox location, which is where I go to do my daily processing. From here, these various captures are tagged and moved to relevant notebooks – which ultimately function no different than folders would in similar applications. Using automations in Zapier and IFTTT, I can create Todoist tasks from any item in Evernote that I decide is actionable.

Regarding "notes", my workflow has an extra level of complexity due to my specific habits and needs. I utilize OneNote in addition to Evernote. For me, OneNote acts as my digital whiteboard and note space where I organize myself as I am performing my work. In a normal day, I create/update a handful of pages in OneNote. As part of my daily review, these notes are tagged and renamed to indicate their contents, date, and purpose. As with Evernote, I use automation to easily send items in OneNote to my task manager and other tools.

The most crucial part of my daily processing revolves around my task manager. I have used many competing services over more than a decade and have found Todoist to be the best for my needs – in no small part due to the powerful filtering it supports. This filtering is vital to my workflow as I commonly create more than a dozen tasks in Todoist every day. I utilize a modified Eisenhower matrix for my tasks to determine priority. I use labels/tags to define task types and context. In my case, these contexts typically define the type of action associated such as phone calls, spreadsheets, or manual labor. Todoist supports Kanban views within projects which I employ to create the time dimension for tasks. Each Kanban column denotes my intended timeframe to work on the tasks within the current week, next week, this month, this quarter, or unknown. The ultimate goals of my task manager process is to distill my task list to a unified "active" list denoting what I should be working on presently. From this active list, I assign dates to my tasks for when I intend to complete them. I group and filter my Todoist tasks by project, area of responsibility, and work type.

The final part of my daily review and processing involves my Google Calendar. As with similar services, Google Calendar offers quick scheduling and editing of events – especially for time blocking, which is my preferred method for time management. I use a hybrid approach in how I delineate my time blocks. For the most part, my time blocks are based on the project or context of my work. For instance, the time I plan out for mundane general work like emails and is set as "clerical" while work planned for a specific project, like "productivity book" is blocked as such. I also color-code the time blocks based on the focus level I intend to utilize – I can handle interruptions during clerical work time or home upkeep but make sure to close off my writing time from interruption. This also allows me set barriers to others about my availability ahead of time. Since I have already prioritized my tasks for each project/subject on my plate, I do not need to schedule individual tasks on my calendar – the time block dictates the broad context of my work and so I refer to Todoist for the specifics. I have found that blocking time by specific tasks creates inefficient micromanagement and inflexibility - where a single disruption can throw your entire schedule into chaos. However, there are times when I have tasks of certain imperative that they must be singularly plotted on my calendar. For this purpose, I have Todoist integrated with my calendar wherein any task with a specific "do" time is automatically added to the calendar. Such flexibility is the principal reason why I adopt the hybrid approach to time blocking.

Upon completing my nightly planning and review ritual, I have my next day completely planned. I generally do not plan out more than two or three days in advance – my weeks have enough unexpected variation that I frequently end up shuffling my schedule anyway. At most, I assign placeholder time blocks for high-priority work that I know will fall on those days. The result of my planning session is a complete picture of the next day and precisely how my time will be used. This means I can start every day and get right to work on creating results rather than planning them. I also get to start the day with clear inboxes in my email, repository, and notebook. Nothing disrupts a day more than working out a backlog in the morning.

It might seem that the preceding processes take a great deal of time, but that is not the case. I have my systems setup in such a way that I can quickly assess and take appropriate action on each item. I have a clear idea about what I determine to be important, urgent, actionable, or irrelevant. This comes partly from experience and repetition and partly from deliberate planning. In most cases, I rarely spend more than ten minutes on my entire daily review. The time I save in doing this at the end of my day is immeasurable – the energy I have available during peak work hours is thus freed up to do actual work tasks rather than planning.

The Daily Grind

With each day pre-planned the night before, most of my waking energy can be directed toward doing work and staying true to my routines. As with so many people, I begin my day by checking my smartphone. However, I am very deliberate in what I am doing on my phone every time I use it. My first order of business is to log my sleep time into a tracker app – in my case I use Google Fit. Next is a brief check of my email. I do only a cursory look at my inbox specifically for any particularly urgent messages. To streamline this step, I have filters setup in Gmail to automatically label and/or star such messages so they come directly to my attention. I make it a point to not send or reply to any emails first thing in the morning unless it is truly urgent. I am very strict in only spending a minute or two reviewing my emails in this manner.

I typically plot my day across four or five blocks of time. The first block of time I ordinarily schedule is an hour for "clerical" work, also known as "light" or "shallow" work. The second block is a focused "deep work" session of three hours. My third time block is full hour in the middle of my day for walking my dogs and a meditation session. Block four is another three-hour session of focused work. Finally, I leave an hour at the end of my workday for more light work and winding down.

The first "clerical" hour of my day tends to be busy with necessary but low-value tasks. This is where I do a deeper dive into my emails and process any new mail that has come in overnight. Unless I am waiting on an urgent email message, I do not check my mail again until my evening review. The next part of my typical "light" work is via a service called Feedly, a powerful RSS, news, and content aggregator. In my line of work, staying current with industry developments is a vital necessity, and so using an aggregator to "catch" this information is incredibly helpful. As a rule, I never blindly browse news or any other general content – it is an unfocused, poor use of time. With Feedly, I can configure powerful filters and search queries that only present news and information relevant to my needs. At this time, I will only read articles that are of urgent need of my attention – everything else is either glossed over or flagged for reading later. For this, I use a reading service called Pocket, that integrates seamlessly with Feedly. Pocket is where I store articles that I wish to take the time to read but are not necessarily relevant enough to my future work to store in my Evernote. For articles that do have such relevance, I clip them directly into Evernote via an automation.

I front-load my day with shallow clerical work to free up the rest of my workday for more important and focused work. My first "deep work" session typically lasts for three hours after a ten-minute prep – bathroom break, fresh coffee, etc. During this session, I mute my phone and move it out-of-sight and disable notifications on my computer. I also employ a program that blocks distracting websites and programs from being opened on my computer during this focused work session. The only tools and programs I allow myself to have open during this time are those directly relevant to the work I am performing.

I take special care not to skip the downtime I schedule in-between my focused work sessions. The reasoning is two-fold. First, deep focus can only be maintained for so long before mental fatigue sets in, and the efficiency of your thinking breaks down. Second, sitting or standing in one place/position for extended periods of time is not good for your overall health. Breaks are important even in the middle of a busy workday. Beware the peer-pressure that glorifies extended periods of grueling work. Disciplined rest is not the purview of the lazy – it is an essential component in maintaining peak productivity.

To further maximize the value of my midday break, I employ a twenty-minute meditation session. Personally, I advocate strongly for meditation practice for everyone - even if only to give the mind a brief break from being inundated with information. Depending on the weather, I might even do a walking meditation. This practice has done wonder not only for my productivity, but also helps manage my latent anxiety.

The end of my midday break spells the beginning of my second deep work session. I take the same measures as before in cutting off potential interruptions during this time. For me, this second daily session of focused work is my most fruitful as I tend to have more energy in the latter half of the day. For this reason, I generally schedule my most important or intricate work in this time block.

My workday ends with another hour-long session for light work. I do a quick assessment of my emails for anything requiring immediate action. Next, I move to the principal purpose of this time block – to cleanup my workspace, both virtual and physical. As I tend to generate many computer files in a day, it is important that I ensure everything is named properly and placed in appropriate directories. The same thing goes for any notes or whiteboard assets I created that day. I then close out my day with a thorough cleanup of my physical desk and work area. This will all ensure that the start of my next workday will not be burdened with my morning energy wasted on cleanup. Laziness today can easily create problems for tomorrow. Avoid that.

Though my day's work is largely completed, I still schedule time blocks for the rest of my day. I do this for two reasons. First, as the cook in my family, I need to allot adequate time for preparing dinner for the day. Second, I understand the importance leisure and family time has for mental health, and so I deliberately schedule it into my calendar.

The final part of my day is the daily review as described previously. The reason I choose to schedule this process so late in the evening is that I regard it as a sort of "closure" for my entire day. This late review is a personal preference in that it allays my anxious nature and gives me mental permission to relax knowing that the day is truly over – the time for worry has passed.

Weekly Planning

My Sunday night review is especially important as it is when I plot out my upcoming week. A single week is the optimal timeframe to make meaningful strategic and tactical plans. To this end, my task manager setup is built around what I plan to achieve in the coming work week. Hence, most of my weekly planning revolves around my task manager to determine which tasks will be on the upcoming weeks' schedule and which get deferred to next week or longer. I use four general categories for all my tasks: Active, Next, Pending, and Deferred. Tasks in my Active category denote what I will be working on in the upcoming week. Tasks denoted as Next are ones that *should*, if things go to plan, be Active next week. Pending tasks are those that I expect to get to in the next two-to-four weeks (end of the month). Tasks that I categorize as Deferred are ones that have no timeframe to get done. Every week, I review each category and promote or demote along this timescale. My process in making these determinations involve my prioritization matrix and other considerations.

Monthly Check-In

I view the end of every month is an ideal time to check-in with how well I am moving toward my bigger goals. I take some time on the last Sunday of each month to assess my progression on all my projects and objectives. It is during these evaluations where I can identify where things are going well or not. Unless there are serious issues that need to be corrected, I do not use this time to make radical directional changes – I save those for quarterly assessments. This is primarily because I wish to avoid the pitfall of dedicating an inordinate amount of time tweaking my systems rather than actually achieving goals. Thus, I view the monthly evaluation as a proverbial break to tie your shoe while on a hiking trail.

The Quarterly Review

No matter how carefully curated a system is, changes in objectives and priorities are bound to happen. In addition, bloat will inevitably set in the longer you use a system. To that end, I utilize a quarterly "super review" to address this as a sort of soft reset of my system. By refreshing my system regularly, I can ensure it remains as elegant as possible with minimal redundancies. These deep reviews help me make sense of what I have on my plate. It is also where I can reassess my current goals and determine if they need revision or redirection. This important process focuses on my task manager, project manager, and repository.

I use Notion as my project manager, which is where I plot out my work projects. My quarterly review assesses progress in these projects and how well my execution fared compared to my planning. Sometimes I find myself making major revisions to my project plans – even outright canceling those that are not going well or have become irrelevant. These instances provide valuable lessons I try to take to heart so I can improve as a project planner.

The part of a productivity system most prone to bloat is the task manager. My objective in reviewing my task manager is twofold. The first is to identify tasks that should either be escalated toward completion or deleted outright. This is another reason why I prefer Todoist as my task tool because it has robust filtering that lets me quickly isolate tasks by age and priority. Every task older than three months has their priority levels reset and are re-processed as if they were new. I often end up deleting dozens of tasks every quarter. I always make backups of these tasks and do a quick review of previously deleted tasks in case I find the rare task that should be reinstated. The other objective in this review is to assess how well my task organization and prioritization routines are performing and make appropriate improvements.

The final part of my quarterly review involves a thorough review of my information repository stored in Evernote. Using tags, I identify which items are relevant to my current and upcoming work. This process also forces me to assess everything I have captured in the previous months. This review doubles as a brainstorming session to discover new projects, ideas, and insights. Evernote has a scheduled reminder feature that I utilize for items that I determine are worth a deeper look at a date/time of my choosing. This is not unlike the famed "someday, maybe" process in the GTD system. Given that I capture hundreds of items into Evernote every quarter, I also make sure to delete as many unnecessary and irrelevant items as possible.

The Annual Purge

Sometimes, you just need to reset everything and start over. This is my mindset every year during the bizarre week that happens between Christmas and New Year's. I have an annual tradition to do a complete reset and reassessment of my entire digital system. Throughout the year, I resist the urge to redesign my systems – it is far too easy to waste many hours doing constant rebuilds of a system when you could be completing actual work objectives. However, by the time the end of the year rolls around I am keenly aware of what does and does not work in my systems. In the final week of the year, my sole focus is a full reset and rebuild of my systems. In some years, they change very little and require only modest tweaks. There have been times where I had to completely scrap my system and start over with completely new paradigms. The result of this process is that I start every new year armed with a (hopefully) better way of doing things. This reset also helps augment the mental boost the new year always brings.

Leave Nothing Undone or Ignored

Chapter 10: Turning Inputs into Actions

All work starts with inputs. Inputs are anything and everything that comes your way in life that needs to have any sort of action performed on it. Ideas, emails, texts, assignments, decisions, shopping lists, requests, orders, and everything in-between is an input. Your entire productivity system begins with these inputs. From there, your processes move inputs along a pathway ultimately ending with each input being completed. The quality of your processes determines the efficacy of your overall productivity.

The principal operation of a productivity system is to receive various work inputs, evaluate them, organize them, determine the right action to take on them, and move them toward to completion. The blueprint for doing this is dictated by your procedures, which involve five steps for your work inputs to move along. These steps are **Capture**, **Parse**, **Prioritize**, **Plan**, and **Action**. These steps comprise the fundamentals of a productive workflow. Return to this concept whenever you feel overwhelmed or lost when planning or overhauling your system.

Stage 1 - Capture

A Productivity system is meaningless without a mechanism to receive inputs – a process known as "**capturing**". Recall that inputs can originate in nearly any format – verbal, visual, written, thought, etc. from any number of sources. Thus, this requires that your processes be versatile enough to accommodate this.

The principal goals of capturing are twofold. First, you want to get every input recorded somewhere external to your own memory. This ensures you do not run the risk of forgetting anything and frees your mind from having to worry about it. Second, capturing collects these myriad inputs into one staging area where they are ready to be processed through the system. A key principle of good productivity is having a complete grasp on what you have on your plate without anything slipping through the cracks.

It is essential to capture anything and everything – from clearly actionable tasks all the way to incidental thoughts and insights. In this sense, everything is actionable – even seemingly isolated ideas. As we will cover later, utilizing your system to gather ideas will ultimately create a space for you to extract immense value from them in the long-term. This is an ability that can bridge the gap from an average output to outstanding output and creativity. It all begins with capturing everything.

In terms of efficiency and efficacy, capture must be focused on recording the task or idea. It should require as little effort and as few steps as possible. As such, the capture stage is particularly reliant on your tools. You want to leverage your tools in a way that every type of input you have can easily be recorded for later review.

The first step in creating an effective capture setup is to identify the vectors through which inputs come your way. For most people, email is one of the biggest sources of work items. You will want to have a tool setup that will capture and/or flag specific emails for later review and action. The same holds true for other communications channels you frequent, such as chat clients. Event and meeting scheduling are other common input vectors. Fortunately, practically every email client has embedded ways to get these synced to any digital calendar. Most of us also utilize research in our work, and much of it comes from various online sources. In recent years, most browsers and devices have built in "share" hooks that allow for rapid recording of online content to external tools. Such functionality enables easy capture of virtually any digital content and is an essential tool for capture.

Perhaps the most ubiquitous inputs in of our lives are ideas and incidental tasks. In days past, the trusty notepad and pencil were the principal tool for capturing these items. Today, there are many technological analogues to the notepad that, when optimally utilized, are faster and more detailed than previously imaginable. The predominant tools for these come from the bustling digital notetaking industry. Most (not all) of these products are platform-agnostic and offer both full-featured desktop versions as well as convenient mobile iterations. A good notebook tool will cover nearly niche in the inputs you need to capture – from emails to meeting notes to brainstorming to quick ideas you just want to record. Given that your digital notebook tool will be where you take the next steps on your captures, this tool should be one that you take great care in choosing.

Stage 2 - Parse

Once you are regularly capturing all your inputs, you will be amassing a sizable list of tasks, ideas, content, etc. The next order of business is to begin organizing and making sense of all this stuff. This brings us to the second stage of the productivity process – **parsing** your inputs. This is where inputs are consolidated, translated, and categorized. Once the parsing stage is completed, your inputs will be ready to be evaluated against each other and your productivity goals. Depending on what tools you are using for capturing inputs, you might have more than one capture list. Ideally, you will be utilizing as few tools and lists as possible to avoid needless complication to your system.

In every place where you capture inputs, you should have an "inbox" location where all inputs initially await parsing. Most task managers have inboxes built right in. Digital notebooks have numerous ways of creating the same effect and usually have settings to direct all new entries to such a default location. Doing this allows you to process your captures in a manner like how you process emails and traditional inboxes.

The first order of business in parsing captured inputs is to determine if they are ultimately a **task**, a **note**, or an **event**. This is the first, highest-order level of this process. In many – perhaps most – cases, you can make this determination at the same time you are capturing. An event is obvious and can be scheduled directly into your calendar. If something is plainly actionable, then it can be sent directly into your task manager. Inputs that are neither events or actionable are relegated as notes. There will inevitably be times when you must capture an input without having the time or immediate insight as to whether it is an event, task, or note. In such a case, be sure to have an input list that is your default for any such outliers.

Determining if an input is an event, task, or note is the most crucial step in the capture/parse process. They are so distinct in purpose that they are parsed in different ways. Recall that a task is an input that is actionable, while a note is non-actionable (at least not overtly or immediately) but important enough for future reference. Events are more one-and-done in terms of processing. Either you schedule them on your calendar, or you do not. Next, we will detail the next steps for parsing tasks and notes.

Parsing Tasks

Inputs get categorized as tasks because they are directly actionable. These should all be captured directly into the inbox of your task manager. You should be processing your task inbox at least once per day – ideally as part of your daily review. Before you can prioritize or plan action on your tasks, they must be "cleaned up" and listed in a uniform manner.

Every task must be written in a way in which its purpose and action is clear and unambiguous. The best practice way to do this is to have the first word in the task name describe the action – typically a verb. Some examples: CALL boss; CREATE spreadsheet; EMAIL client; INVESTIGATE claim; SCHEDULE meeting; and BRAINSTORM ad campaign. In cases in which your capture action is to write directly to your task list, then you can do this step automatically. For tasks that are not initially formatted this way, you should consider rewriting these so that they conform to this standard.

The next aspect of parsing out your task list is to assess their contextual dimension. This can include many categories ranging from type of work to which project or area of responsibility to which it applies to the nature of its deliverable. Most task managers have tag/label features along with folders and sub-folders. It ultimately comes down to your preferences which to utilize for your contexts. I recommend you use tags/labels to describe the general type of work and deliverables of your tasks and folders to group them by project or area of responsibility.

For example, say you must do a sales analysis for Product X which entails you using a spreadsheet. Your task "*ANALYZE Sales for Product X*" will be tagged with "spreadsheet" and "report". If Product X is part of a specific project, then it will be placed in a specific "*ProductX*" folder/list in your task manager. If it is not a project and the report is part of your day-to-day work, it can be placed in a general "work" folder/list. If reporting of this sort is a major part of your job, you might consider an area-of-responsibility folder/list for reporting. Your folder views dictate everything involved with Product X while the tags allow you to find and sort all your tasks by work type and deliverable. This will allow you to do useful things like get a quick list of all the spreadsheet work you have on your plate or how many reports you must deliver.

When you are finished with the parsing stage of your tasks, the inbox of your task manager should be completely empty. The inbox should strictly exist as a temporary holding area for unprocessed tasks. Everything should ultimately reside in a folder, grouped under the context most appropriate for your work style.

Stage 3 – Prioritize

A to-do list is ineffective if you don't know which tasks take priority over others. You must be able to quickly determine what you should be working on first and what can wait until later. This is where the all-important process of prioritization comes in. If there ever was a single most-essential skill for effective productivity, the ability to properly prioritize work might be it.

Ultimately, the priority of a task comes down to the ramifications of its completion or non-completion. Despite the numerous approaches to prioritization out there, there are only two aspects of a task that matter: urgency and impact. Urgency is self-evident in that there is a specific window of time that exists in which the task can be completed. Beyond that time, completion of the task is either impractical or impossible. The impact of a task concerns how its completion (or lack thereof) affects other tasks and/or outcomes – a task that is a prerequisite for other tasks often carries more impact than one that does not. Impact also denotes the overall value derived from its completion. Impactful tasks are important. Tasks with little or no impact are not.

As we see from the above, you only need to make two determinations about a task to assess its priority: if the task is urgent and if the task is important. This is the essence of the famed Eisenhower Matrix – the only prioritization methodology I recommend. It is powerful, elegant, and remarkably simple to implement. It is about as ideal a process as you can find. In this framework, every task has but two binary dimensions that determine priority: urgent or not urgent and important or not important. In determining each dimension, you are left with four distinct levels of priority for every task: important and urgent; important and not urgent; unimportant and urgent; and unimportant and nonurgent. The pervasiveness of this matrix is why so many task management tools offer at least four levels in their priority settings.

While the first and last priority level are self-evident, the middle two (unimportant/urgent and important/nonurgent) might depend on personal preference. The original concept of the Eisenhower matrix was built around a workflow where work tasks can be delegated to others – a luxury many of us do not have. The second highest priority level is typically important/nonurgent as recommended by productivity experts. It comes down to personal preference on which side of the urgency vs importance you want to fall. Neither approach really gives up anything in terms of defining task urgency or impact. However, by placing greater priority on important tasks vs urgent tasks, you are taking a step toward overcoming a psychological tendency known as "present bias" – where we tend to focus on short-term achievements at the expense of long-term goals.

For such a decision matrix to be effective, it needs to be consistent, and so you must be very clear on what your criteria are for determining the importance and urgency of your tasks. This is achieved by asking a series of unequivocally yes/no questions for each task. These questions should be explicit so that a "yes" answer to any one of them will immediately indicate if the task is important and/or urgent. As every person is different, you will have to determine these questions for yourself and your unique needs. To get you started, below there are two sets of example questions regarding task importance and urgency that are common to most people.

Is this task important?

1. Is there an emergency?
2. Will completing this task bring high value results?
3. Does this contribute to my overall life goals?

4. Does this contribute to my larger work goals?
5. Is this task a prerequisite to other important tasks?
6. Is this task a prerequisite to numerous other tasks?
7. Will not doing this create negative consequences?

Is this task urgent?

1. Is there a specific deadline to complete this?
2. Is there a finite timeframe to complete this?
3. Is somebody waiting on me to do this?
4. Will not doing this create additional work?

Once every task has been evaluated with these questions, each will fall into one of four categories of priority:

Priority 1: Important AND Urgent

Priority 2: Important AND Nonurgent

Priority 3: Unimportant AND Urgent

Priority 4: Unimportant AND Nonurgent

Stage 4 – Plan

At this point your task list should be concisely written, categorized, labeled, and clearly prioritized. This brings us to the planning stage of the process, where you are plotting out how you will use your time. As this is the time-planning stage, this will be done primarily during your weekly review with (hopefully) minor tweaks during your daily reviews. This process is an essential component of good time management.

The first stage of planning out your weekly time usage is condensing your task list into a list of what you specifically intend to work on in the coming week. If your task management tool supports Kanban views, this is a great context in which to utilize it. If not, the same function can be performed via labels/tags. I use the term "active" to denote my current weeks task list. It is here where the fruits of your task prioritization are evident. In most cases, priority one tasks (important/urgent) will need to be done immediately, and so will be the first to make it onto your active list. From here, your discretion will be the key to which lower-priority tasks will make it into your active list. Though the principal objective in this process is to determine your active tasks for the week, your secondary objective is to group the remaining tasks based on when you expect to get around to them. At the very minimum, you should have a "next week" categorization. I recommend at least a "this month" category as well. With each weekly planning session, you should be moving these tasks along this time vector toward your active status. It is for this reason it is recommended you utilize a tool with Kanban functionality.

When finished, this process will result in a task list specific to what you will be working on in the present week. From here, you can plan the specific times in which you will be completing them. A most effective way of doing this is through a technique broadly known as time blocking. This is primarily effectuated via your digital calendar tool and its event scheduling function. Time blocking is just what it sounds like – scheduling time on your calendar for specific activities or tasks. In the same way you would schedule a meeting or doctor visit, you schedule a block of time to do work.

Time blocking is particularly effective for productivity in that it sets boundaries. As it is expressed on your calendar, it creates a clear, visual representation of what your time and attention will be specifically devoted to. This visualization also helps you get an accurate idea of how much time you have in your day and how it is utilized. Since work is scheduled on your calendar, it is a great aid against procrastination - just as you don't linger on a client meeting or zoom call, you don't put off your allocated work. There is specific time allotted for specific activities to be performed. This makes it easier to focus on what you are doing and shut out distractions – if it is not pertinent to the time block you scheduled, then can be ignored. It also lets you stave off interruptions in the same way a scheduled meeting does.

From a Psychological standpoint, planning a distinct time to do specific work makes it easier to focus on that work alone.

There are four general approaches to time blocking. The first is known as **batching**, where tasks of similar work types are grouped together to be completed in the same block. This method is great for attacking groups of smaller tasks and lets you stay in one mental "mode" in getting them done.

The next approach is to strategically block time based on a specific project or work responsibility. For instance, you schedule a three-hour block of time to work on all tasks pertaining to Project X regardless of the specific types of work each task entails.

The third approach is scheduling a specific task directly to your calendar. This is best utilized for time-intensive, complex tasks or ones with specific timeframes – you do not want to clutter your calendar with dozens of tasks. Most task manager tools support calendar integrations, which make this type of time blocking easy to do.

The final approach is the broadest in that it effectively blocks out the entire workday: **day theming**. With day theming, the entire workday is scheduled to be devoted to a specific type of work or project. The theme of the day's work is as broad or specific as the context scheduled. For instance, Monday might be specifically set aside for Project X, while Tuesday is reserved for more broadly as a "writing" day.

Fortunately, all four approaches to time blocking can be employed in a hybrid manner. As with many aspects of your digital system, flexibility is often worth it – an overly rigid system will always fail in the complexity and chaos of real life. Each approach to time blocking works best for certain types of work and focus. Redundant or routine tasks are best planned using task batching. Work relating to a specific subject or project is best served by strategic time blocking. Critical tasks or tasks tied to a specific timeframe benefit from direct scheduling on your calendar. Larger-scale work themes or projects often benefit from being the sole focus of a workday, and so day theming is appropriate for them.

As seen above, a hybrid approach to time blocking allows you to accommodate virtually every type of work you might encounter. Since every method utilizes your calendar in similar ways, they can coexist there without creating undue confusion. You can have a calendar with Monday day-themed for writing; Tuesday split between time-batched shallow work and work on Project X; Thursday blocked in the morning for meetings with a scheduled task "sign up for conference when registration opens" at 3PM; and so on. One further recommendation is to use different colors for the different time-block types on your calendar. For instance, I use dark red for my regular calendar events, gray for batched-work blocks, pink for my strategic blocks, yellow for day-themed blocks, and pink for scheduled tasks.

Stage 5 – Action

One of the most dangerous pitfalls of productivity systems is that they can create false sense of accomplishment. Regardless of how organized your task list or curated your information, it is ineffectual unless you are accomplishing meaningful objectives. It is easy to be seduced by the feeling that you have achieved anything worthwhile when all you did was organize and plan. Planning and organization are vital prerequisites to productive work, but it is not work itself.

In the process of doing work, the action stage is principally a function of behavior. As with every other aspect of productivity and organization, it is a set of skills and habits that can be cultivated. Productive behavior is a confluence of four principal aspects. The first is your **mindset** and how your attitude fosters (or hinders) your efforts. Next is your **environment** and how it influences your ability to work. The third involves your **body and health** – an often overlooked yet critical component of efficiency. Finally, there are your **habits** which comprise the automatic, effortless behaviors that we program into ourselves consciously and unconsciously. All of these will be discussed later in further detail.

Organization and planning are the steps taken to get to this point - where work starts to get done. The reason we put the time and effort into capturing, parsing, prioritizing, and planning is to remove as many barriers to action as possible. It is far easier to focus on completing meaningful work when you have already organized and planned what you have on your plate. This frees up your mental and behavioral resources for the work at hand without having your bandwidth diverted. Keep this in mind when you are processing your inputs and tasks – you are creating clarity for your brain to work more efficiently when the time comes.

All Work Exists on A Continuum

Chapter 11: About Project Management

Every productivity system is a project management system at heart. They share the many of the same concepts and purposes: the curation of relevant information and the identification and organization of action steps in service of achieving greater goals. In your journey to understand the essences of productivity and organization, you are also learning the fundamentals of project management. Sometimes, you will encounter a project or goal that needs more direct planning than what can be handled with a notebook and task manager. In cases like this, you might benefit from direct project management to properly tackle them.

Though the intricacies of project management are beyond the scope of this book, its fundamentals are more than sufficient at the personal scale. The first consideration is to understand what constitutes a project in the first place. A project consists of a specific goal as achieved through a sequence of steps. This goal is measurable and constitutes a true finished endpoint – a recurring process or routine is not a project. The project goal produces a result that is unique and non-repeatable. Projects have clear beginning and ending points and are constrained within a finite timescale. Writing a book constitutes a project, as does building a website. Manufacturing a good or running a routine analysis are not.

A project is a plan comprised of sequential steps ending with a clear deliverable. You begin by determining if anything in your workload qualifies as a project. Next, you will describe in specific terms what the goal is and by what criteria will you know it has been accomplished. From here, you need to determine every major step must be taken to achieve the goal. These steps must be sequential in that one must precede the other as the project advances. Do not worry too much about the smaller steps here - at this stage, you are looking for what are known as milestones. Milestones are smaller deliverables within a project that comprise the larger goal. For instance, in writing a book, your milestones would be something like “research subject X” or “write first draft of part one”. You will establish the more miniscule tasks that lead up to these milestones later.

Once you have established what your milestones are, you should have a clear idea about the size and scope of your project. This is where you can make a meaningful estimate about how much time the project will take for you to complete and the extent of the work involved. Such information is invaluable for figuring how it fits into your current work priorities. Knowing the size and scope of your projects is half the battle in figuring where your efforts are focused.

The final stage in this process is creating specific tasks from your project milestones. Professional project managers employ a methodology called **Work Breakdown Structure** (WBS). Essentially, this is the practice in which milestones are progressively broken down into smaller component lists of tasks. An example: the first milestone in my book writing project is to "write first draft" and so my first sub-objective is "write first draft of part one" and my final subtasks here are "write chapter one", "write chapter two", and so on. By continuing to break down my project in this way, I will have generated a full accounting of the tasks that must be completed to fulfill my project goals.

The project management process above can be performed using different tools. Even something as simple as a text document or spreadsheet can get the job done. If your work frequently involves project planning, then you might benefit from a specialized tool like Notion, or Trello. Consult the earlier chapter regarding project manager tools for ideas and recommendations.

This stage results in a complete list of tasks that comprise the entirety of the project. From here, you can input and process these tasks into your task manager as normal. It is highly recommended that you utilize a dedicated list/folder in your task management tool for each project. This allows you to easily visualize and prioritize the work in your project separate from your other tasks.

I want to close out this section with a caveat about professional project management. It is a fascinating and complex subject, and I only scratched the surface on its intricacies. The information I presented here was greatly simplified into the purest concepts. Though this abridged approach is useful in and of itself, it is not representative of professional project management nor necessarily its best practices. If you find these concepts to be of real use, I highly recommend digging deeper into the subject. You will not be disappointed.

Fail to Plan = Plan to Fail

Chapter 12: Organizing Information

The ability to manage digital information has become an increasingly critical life skill that will only become more important in the future. As we have seen, information comes at us in many different forms and contexts. To thrive in the digital world, we must take deliberate agency over what information we let into our lives and be proactive about what we do with it. This starts with how we regard information – everything relevant is viewed as an input to be filtered out or captured and processed. With this mindset, it is easier to see what information has any potential value to us or not. Make it a habit to look for the value in whatever information comes your way. Make it a habit to swiftly reject any such items that have no value to your life. This is an important practice to cultivate, and in doing so made a first step in filtering the flood of information we all encounter daily.

Recall that every captured input can be broadly categorized as an event, note, or task. There is an optional fourth category you might wish to utilize: "**disposable**". Disposable captures can be thought of as analogous to daily newspapers or the ten cent comics of days gone. This is media or other content that you want to consume later but is not important enough to retain long-term. Your repository tool could be used for this purpose, with such items being flagged as disposable. The main caveat here is that you will need to regularly flush out and delete this content periodically or when you consume it. My ultimate recommendation here is to maintain your disposable captures in a wholly external location than your repository or notebook. Services like Pocket and its competitors enable easy capturing and recall for text and multimedia alike.

Categories, Projects, and Areas of Responsibility

We all live multi-faceted lives wherein our attention and energies must be distributed across different priorities. If we are to exert any real mastery over our lives, we must have a clear picture of what these facets entail. It begins with a question: what are the different aspects and responsibilities that comprise my life? In the parlance of productivity, business organization, or project management you might also see this referred to as "**Areas of Responsibility**". These areas broadly cover our responsibilities and priorities. For instance, I break my life down into four life categories: Personal, Home, Hobbies, and Work. They are all self-explanatory in their names (I regard upkeep of my home a distinct element of my life), and I can neatly fit every task, activity, or piece of information in my life into these boxes. Undoubtedly, how you categorize your life will be different.

Every component of your personal organization plan should use these categories as the basis for how they are organized. This ensures unity across the various tools that you utilize. It also further reinforces the habit to think of your life and responsibilities in this way. This type of clarity on how your life is structured is an essential part of productivity and organization.

Be careful not to confuse projects with your life categories. A project has a distinct deliverable and finite period in which it is created. Categories are broad and cover the largest aspects of your overall life. Projects exist *within* your categories. As an example, remodeling my guest bathroom is a project that falls under my "Home" category.

In plotting out your organizational system, begin by establishing your life categories, and use them as the basis for how your tools are configured. Your email folders, task categories, notebook stacks, etc. should all be setup using this categorization scheme. Specific projects should be nested within their relevant categories as well. The goal here is that every part of your digital system should be using the same category schema. This is essential in creating clarity and uniformity in your systems. It also allows you to introduce new tools or complexity (if needed) into your system more easily – you already have a predetermined way of setting things up.

Organizing Email

Email continues to be one of the biggest sources of bulk information for many people. It is a principal pain point for overwhelmed workers the world over. Hence, taking control of your email is one of the most important things you can do for yourself and your productivity.

It is essential to understand that email is more than a communication channel. Today, email functions as an information, file storage, and record-keeping system all at once. Everything from plane tickets to shopping receipts are transmitted and stored there. This is also to say nothing about the countless file attachments that reside in so many of our accounts. Email is a vast trove of information and must be curated and organized as such.

A properly curated email account has four characteristics. First and foremost, useless or inconsequential messages are never retained. Second, all pertinent messages are labeled and/or categorized in a clear manner. Third, the inbox contains as few messages as possible with zero being the goal. Finally, older and obsolete emails are regularly purged from the account.

Delete the Useless

Do not be afraid to delete emails. The only messages you should retain in your account are those that are actionable, informational, or necessary for record-keeping. **Actionable emails** are those to which you would respond or act in accordance with the content within. **Informational emails** contain information that will be needed for future reference. **Record-keeping emails** are things like receipts and memos – not necessarily things you will go back to but to tell a story in aggregate. Organizations commonly find value in retaining emails like this because they are so useful in recollecting decision making processes and justifications. Everything else that comes into your inbox should be promptly deleted. This includes messages that might be worth a read initially but pose zero future value in retaining it. Develop the habit of deleting useless messages outright. This should be the very first thing you do every time you open your email client. In doing so, you are going a long way toward avoiding email overload.

Tags and Categories

The next step in processing an email inbox is to identify the purpose of the remaining messages. They are either messages that require your response, contain actionable content, or contain informational content. Messages that do not fall into these criteria are either useless (which you should have deleted previously) or warrant retention for record purposes. These messages can be archived outright. The others should be labeled as to their purpose – in my case, these labels are "reply", "action", and "info". Most email clients support custom views or quick sorting by such labels. Bear in mind that email clients use different names for this functionality, such as labels, tags, or categories.

Email clients offer slight permutations of folders for grouping messages. In days past, it was important that virtually every message was grouped into folders based on subject or context. The advent of powerful search and filters in email clients has made this step less crucial. If you choose to group your archived messages in folders, be sure that it is functionally necessary and not something that can be effectuated with a saved search or filter.

Inbox Active

The inbox was never intended to be a long-term location for email messages. Treat your inbox as your active space – if the message is retained here, it is because you will need to access or act upon it soon. Everything else gets archived. Through this practice, your inbox presents an at-a-glance grouping of the specific emails on your plate. Once an email is no longer immediately relevant to your near-term work, it can be archived.

This is the essence of the "inbox zero" method of email management. However, I prefer to think more along the lines of "inbox active" – if it is actionable or relevant, it stays into the inbox until it is truly off the table.

Email Retention and Purging

Over time, email accounts can amass an immense number of messages and attachments. This inevitably leads to bloat that can impact email performance and efficiency. The searchability of email can be hindered when there are too many irrelevant messages coming up in queries. In addition, most email providers have size limits to accounts in terms of messages and/or overall file sizes (which include messages and attachments). Then there is the matter of security and privacy – breaches are a part of life in the digital age, and even the best security practices are undefeatable. This brings us to the matter of email retention practices.

One important matter to consider before doing any email deletion is to consider your legal and professional obligations. For nearly every industry, emails are legally regarded as official records and documents. The best-practices for retention times vary between industries –ranging from one to as much as ten years or longer. Some organizations and industries are even subject to rules wherein emails of a certain age *must* be deleted upon reaching a certain age. When in doubt, you are best falling back on the IRS minimum record retention period of seven years. This status of email as official records also should inform how you conduct yourself in their use. When communicating via email, be careful what you say and how you say it. In addition, improper deletion of emails can land you in hot water legally and ethically. Govern yourself accordingly.

Given that retention periods are defined by years, the best time to do an email purge is once per year. This is ideally done as part of your annual review of your entire productivity and organizational system. Bear in mind that most email clients do not outright delete emails, but rather place them in a trash folder. Messages in this trash folder will eventually delete permanently, which is commonly thirty days.

Capturing Emails to your Repository

Though email is invaluable for information storage, it is not optimal for its curation. For this reason, you should be in the habit of capturing certain varieties of emails into your repository. This is especially useful for email containing receipts, legal info, account info, etc. – anything that includes something you need to recall frequently or curate with other snippets of information and notes. Most leading note or repository services have functionality that lets you easily capture emails – either by forwarding the message to a special address or with the click of a button. Automation services provide further ways to do this, including using tags or full automated capture of emails based on certain criteria.

Notes, Clips, and Bits

The entire point of capturing information to your repository is so it can be accessed for future recollection and reference. This likely is the part of your system that will amass the largest volume of inputs including notes, articles, documents, and other clippings. Proper organization of this information is therefore critically important.

Parsing notes is done in a similar manner as tasks, though your main concern is that they are ultimately easy to search and browse for later recall. Since notes are not actionable, you do not need to name your notes with the same degree of specificity as you would a task. As with tasks, tags/labels are best utilized to describe the type of content a note has. Example tags include @article, @idea, @screenshot, @handwritten, @photo, @pdf, or @scratch. Notice how these examples all utilize a special character ahead of them. This special character identifies the type of tag they are. It helps to differentiate them from tags that identify content and context for easier search and curation.

Tagging is especially important for notes captured in your repository tool. It adds an extra dimension of searchability to your entries beyond searching functions. Because repository notes are likely to be numerous, the more searchable a note is, the better. Therefore, it is worth the time to assign proper tags as you process and parse notes in your inbox.

Most notebooks support folders (or equivalents), and so those should be your principal way to group them by project or life category. Remember to use the same category structure as the rest of your system. If done with proper diligence, the captures in your repository should be easy to search and sort as you need them.

Since notes are not actionable, they require no further processing once they are sorted and tagged. Of course, it is possible for a note to become actionable in the future. For this reason, it is vital to have your notebook and repository tool connected to your task manager. Depending on your tools of choice, this can be done natively in the app or through a third-party automation service. Setups like this typically allow you to create tasks in your task manager linked to your note item in question. One such example would be using a service like IFTTT that will create a task in Todoist whenever you apply a specific tag such as #sendtotodoist. The useful thing here is that the task will link directly to the note you assigned the tag to, and all you need to do is parse your new task like you would any other.

One final part of utilizing an information repository is curation – pruning irrelevant entries and elevating the germane. You should be doing regular reviews of your repository at least a few times per year. These reviews can be regarded as half cleanup and half brainstorming: The inspiration for your next project or new insight into current work could be lurking within your trove of notes. Notes are also referenced every time you refer to them as part of doing your work. Regardless of when you are reviewing them, you will need a way to flag notes as being relevant or irrelevant. This is also best effectuated using tags. In this context, notes only have one of three statuses: **useful**, **irrelevant**, or **neutral**. I utilize two special tags for marking these distinctions: one called "!float" and the other "!sink". The choose this nomenclature because it indicates that I want to "float" relevant information to the top and "sink" the irrelevant to the bottom. Content I float to the top is relevant to what I am working on or want to circle back on soon. The sink tag indicates I have determined it has little value or relevance at all. Notes with neither tag are regarded as neutral. When referencing back to my notes, I filter by the float tag and thus have a condensed selection of notes I have deemed timely and/or relevant to the work at hand. Notes I flag as sink are often ultimately deleted from my repository during my quarterly reviews.

Notice again how I use special characters as prefixes to my tags. This serves an important purpose in how my repository tags are managed. Tags without any special characters are simply subject matter tags that comprise the bulk of my tag list. I use exclamation points ahead of tags that indicate a special status, such as relevance as described above. Tags that indicate the type of note are prefixed with an ampersand. Hashtags precede tags that trigger actions tied to my automation services. By using these prefixes, I can artificially create different categories of tagging for different purposes that are easy to see at-a-glance.

Managing File Systems

An often-overlooked aspect of digital organization is how computer files are managed. Nearly all knowledge work involves creating, accessing, and organizing various types of computer files. Without a concise and consistent naming and directory protocols, it is easy to create a chaotic and confusing mess. Few things waste time like having to search through a jumble of files and folders to find something.

The advent of cloud storage has allowed personal and professional files systems to live independently from devices. If you are primarily syncing your files on a desktop computer, you are best keeping your files under the main directory your cloud storage is installed on. Leading services like Dropbox and OneNote do allow you to sync files and directories outside their home directories. Use discretion in using this feature. For file storage, the more consolidated everything is, the better.

Effective file management is achieved by following a protocol of best practices. Firstly, a file system must be applied consistently and uniformly throughout the entire directory structure. It is counterproductive to use different organizational paradigms in different directories. Second, the best file systems follow a hierarchical structure, in which sub-folders are progressively narrower categories from the larger parent folders. The deeper the folder is in the hierarchy, the more specific the context should be, not the other way around. Folder names should be reflective the contents within, either by context or description. Be careful not to create too many (or too few) levels to your folder hierarchy – using no more than three is the general guideline amongst experts.

Unless dictated by job requirements or other external factors, how you structure your file directories is a function of personal preference. As with every other part of your system, it must serve your needs. A good starting point in figuring this out is to look back to your areas of responsibility. These make ideal top-level directories. Folders containing files related to specific projects can be nested within these relevant folders. Non-project files under these categories can be sorted in folders delineating time periods or generalized sub-categories. Again, you need to experiment and determine what works best for your workflow and file output.

It is advisable you also maintain top-level directories that exist for functional purposes. Examples include a directory for web browser downloads, record-keeping, file backups, etc. Many cloud storage providers offer automatic backups for photos and videos captured on smartphones. In these cases, you will have an automatic directory where these are stored. If you tend to generate or acquire digital items frequently, you might want to consider an inbox directory as well. As with inboxes in your email or task manager, it is a temporary location for files to exist before you process them into their proper directories.

How you name your files is the final component to effective file management. Your file naming conventions must be coherent and consistent. This is especially important for ensuring your files are optimized for search. Though you should strive to maintain uniform naming conventions across all file types, sometimes this is impractical as different formats and contexts require different types of information to be conveyed: a recurring report spreadsheet, a one-off memo, and a version of computer code have different descriptive priorities in how they are to be named. Use the right naming conventions based on file types and/or context but be sure to apply them uniformly.

As with folders and directories, there are best practices to follow in terms of how you name your files. Venture to keep file names as short as possible. The name itself should be descriptive of the file and can contain several descriptive elements. Examples include project names, area of responsibility, version number, work type, location, dates, excerpts, and more. Avoid using spaces by substituting hyphens, underscores, or periods in their place. You can also do this by capitalizing the first letter of every word. Refrain from using special characters. I do make one exception here for sorting purposes – using a character such as an exclamation point at the front of a name is a quick and easy way to quickly float files and folders to the top of lists and searches. If you use dates, format them in the ISO 8601 scheme for optimal search and sorting: YYYYMMDD.

Archives and Backups

Just as with other information storage mediums, there comes a point where older files need to be cycled out of your system. The simplest approach is to simply delete all files (barring special cases) after they reach a certain age. As with emails, you must do so in accordance with whatever retention laws and guidelines that may apply to you. Like emails, cloud storage providers do not fully delete file until a period has passed as they sit in a recoverable "trash" directory. A less drastic method of handling old files is to maintain an archive directory – as files reach a certain age, you simply move them here rather than delete. If you are cognizant of limited cloud storage space or file bloat, you might consider keeping your archive on a different storage service or drive. In my case, I use OneDrive for my active files and Dropbox as my archive. This is admittedly costly, but it means all my files are safe (from deletion, not necessarily security) while my old stuff is not cluttering current files.

As convenient as cloud storage is, you should still strongly consider maintaining backups on your own drives – particularly for important files. This inoculates you from cloud outages, disruptions, and discontinuations. External USB storage is relatively cheap and fast. If you go this route, make it a routine to back up your most important files to an external drive. Get yourself a fireproof lockbox or safe and keep the drive there. If you are particularly concerned about file loss, alternate backups between multiple drives. One caveat: if you are under a legal, professional, or ethical obligation to delete certain files, you will have to do so for your backups as well. This can be a time-consuming process, so bear this in mind when deciding which files you retain in this way.

Data Without Context is Little More Than Junk

Chapter 13: Routines – Keeping the System Going

Where productivity and organization systems are static constructs, the state of actually *being* productive is active and dynamic. On their own, systems are merely steps and protocols by which inputs are sorted and translated into concise plans for action. It is only through repeating action that a system creates any value in this endeavor. These systems create the opportunity for productive work to happen. When properly utilized, you are left with your information organized, your tasks sorted, and your time prioritized. This is the starting point for getting everything done, not the ending.

These systems cannot work unless they are maintained and operated with regularity – every input must make it through to their final state. If you are not actually using the system as intended, then the benefits they impart are meaningless. This is achieved through routines. When it comes to results, routines *are* the system.

The Hierarchy of Routines

Keeping your systems running require regular processes. There is a progressive hierarchy to the focus and purpose of these routines based on the timeframes in which they are performed. Capturing inputs must be done immediately as they come to you – straight into your inbox. These inputs should be processed out of your inbox daily. Perform a planning session before every week. Goals and priorities are reviewed every month. Tasks, objectives, and get thoroughly reviewed each quarter. Finally, all systems are flushed and reset at the end of every year. These are the routines that keep your systems functional and invaluable.

1. Capture constantly
2. Process daily
3. Plan Weekly
4. Review Monthly
5. Strategize Quarterly
6. Revise Annually

Capture Everything

The fundamental point of a productivity system is to ensure everything that comes your way is addressed. This begins by making sure that every one of these inputs enters your system to begin with. You should have ways to capture inputs for every time, situation, location, and context you find yourself in. We have previously covered in-depth the many tools and services that can assist with this. But having the tools and capacity to capture inputs is nothing without making it a habit to utilize them. Make it an obsession to cultivate this practice. The goal is to be able to recognize inputs effortlessly and quickly as you encounter them and capture them to your inbox(es) immediately.

Process Daily

The end of every day is where the success of the next day is planned. Every day should end with time set aside to process all the days' captures and plot how the next day will unfold. Every inbox must be cleared, and any overdue tasks are rescheduled. The work priorities for the next day are decided the night before and all workspaces are cleared. This ensures that mornings are not spent wasted on cleanup from the previous day. As the entire day has already been time blocked previously, it lets you start immediately on getting work done rather than making decisions on what work to do. Every day should begin with a clean slate and a plan in place. Your daily processing and review is the most critical aspect of achieving true productivity.

Plan Weekly

As a block of time, a week is the most basic period for strategizing goals. It is long enough for meaningful milestones to be set, and short enough to gauge efficacy and change tactics accordingly. As with days, every week should be planned out at the very end of the previous one. Your task list should be fully audited at this stage to identify everything that takes priority to be completed or worked on in the coming week. This process should result in a definitive list of every task and objective that is active for the week. This way, everything else not on deck for the week can be ignored. Such pre-sorting makes the daily processing step much easier – the task list is pre-sorted.

Review Monthly

If you are keeping up with your daily and weekly processing routines, the monthly review can be focused on evaluating how things are going. If your system is getting bogged down with overdue tasks or unprocessed inputs, this is a good time to get ahead of them. Taking account of the previous month provides a good view of how your productivity and organization is faring. Measurable progression and results should be evident at this timescale. If things are not progressing well, you need to figure out what was going wrong and how to do better in the next month. For every project and area of responsibility, set clear and measurable goals for the month. Be sure to review these monthly goals at the start of every weekly planning session to help ensure you stay focused on these targets.

Strategize Quarterly

The timescale of the three-month quarter is long enough to leverage long-term strategy and assess system bloat. End every quarter with a thorough assessment of how your productivity and organization are going. This is also a good time to do a deep review of your task lists. Older tasks should be re-evaluated for priority and relevance. Extraneous older tasks should be deleted promptly. If a task has stayed unfulfilled for an extended, why is it even there? This should be part of a larger process of reaffirming what your longer-term objectives are and what your strategies will be to achieve them. Your re-prioritized tasks should reflect your revised strategies.

Each quarter is also a good interval to take a deep dive into your notes and other captured content. This review is different from that you would do in the process of carrying out work. This review should be focused on looking for new ideas and insights. Throwaway ideas, articles, or notes in your repository can often provide the inspiration for new projects or visions. A quarterly dive into this collection of information is when you go in search of this potential treasure. In addition, it is a good time to identify redundant or useless items that can be deleted.

Revise Annually

All systems are artificial and imperfect, consequently requiring regular evaluation and improvement. Productivity enthusiasts commonly fall into a trap wherein they are perpetually assessing and reworking their systems. Regular tweaking of your systems can create a false sense that something is getting accomplished. Such a preoccupation contributes nothing toward achieving your larger goals or purposes. This is counterproductive beyond merely being a waste of time - the efficacy of many systems can only be accurately evaluated after sufficient usage. Naturally, incremental improvements to systems are always warranted if they are addressing a specific failing or inefficiency. Nevertheless, fundamental, sweeping changes to your systems are best reserved for larger timescales as it minimizes disruption to your workflow. Your systems should be rebooted or revised annually. Even the best designed systems will inevitably amass bloat and redundancies.

Take time at the end of every year to make a thorough and honest look at what value your systems provide. In this process, you are looking to identify what works, what doesn't work, and what could work better. Ask critical questions: Does this system make my life easier (not necessarily simpler)? Am I using this system just for the sake of using it? Does it save me time and/or energy? What is my least favorite aspect of using this system? Are there complexities in this system that are unnecessary? Is there anything that can be simplified? Am I using too many tools or services with my system? Are there better tools and services I could be using? Where did my system fail me in the past and why?

Depending on how your evaluation goes, you might only need to make minor changes to your systems. However, do not rule out the possibility of having to throw them all out and start over. All the same, this will be an involved process and should be treated with the appropriate degree of importance. Never forget the fundamental reason you have these systems and processes in the first place: to help you stay organized and productive while maintaining your sanity. The systems exist to serve you and not themselves. In many cases, especially with famous named methods (such as GTD and its cult-like fanbase), users get caught up in serving the designs of the system. Take care to not become a slave to your processes.

PART IV: Behavior

Chapter 14: The Will to Action

There seems to be an epidemic of busyness in our society. Lamentations of days filled with nonstop meetings and endless tasks are common aspects of normal conversation. This busyness is undoubtedly a factor in the proliferation of stress-related mental illness today. This is especially problematic for workers in economies where the increase in workloads has not seen a corresponding increase in compensation. This trend shows no signs of changing, and so the challenge of balancing work productivity and mental health is more critical than ever.

The first step away from busyness toward productivity starts with understanding where it comes from. Busyness is a state of being overwhelmed by responsibilities, deadlines, information, and tasks. It is characterized by a sense of chaos and lack of control over what is going on in life and work. The counterpoise to such a conundrum is purposeful organization. Busyness stems from not knowing what you should be working on and when. Without an organized accounting of what tasks and responsibilities you have before you, everything is on you at once with no sense of direction. Having read this far into the book, you are aware of the tools and techniques needed to organize these factors. However, knowing is only half the battle. Without a deeper sense of your purpose and ultimate goals, being organized just makes you a human automaton.

Most of the literature on productivity and organization marks the contrast between being busy versus being productive, but this is not the entire picture. I spent nearly fifteen years working long hours and little time off with seven-day weeks and all-nighters were common. Despite developing an anxiety disorder and multiple health issues, I managed to achieve many job goals and milestones during this period. By traditional measures, I was highly productive, which meant that I was meeting the definition of a "successful" person – especially as a man, who is expected to pursue this type of achievement above all else. Unfortunately, despite all my deliverables at work, my quality of life was abysmal. I had success, but no fulfilment. My lived experience here is not uncommon, and I expect that readers of books such as this can relate.

The takeaway from the above anecdote is that merely being productive is not enough. People have more resilience than they realize and can maintain high levels of output at great personal cost for extended periods of time. But at what price? Physical workouts are not regularly done to the point of collapse for good reason. The mind and psyche should be treated with the same deference. Traditional definitions of productivity are fixated on output and results, but often ignore our sense of greater purpose and quality-of-life.

Goals and Higher Purpose

Instead of addressing just the contrast between being busy versus being productive, we should be assessing busyness vs productivity vs fulfillment. The solution is that productivity should be redefined to include personal purpose as a critical component. This means that the key to fulfillment and purpose lie not with how productive you are, but what the *goals* of your productivity should be. To turn a metaphor on its side, the destination is as important as the journey.

Achieving goals is important but discovering what those goals are is even greater. We all lead finite lives, and it falls to us to find and create purpose for ourselves. Purpose is the prerequisite to life fulfillment – the very thing that makes our limited existence worth it. Nobody else has the responsibility nor the capacity to create contentment for you – it is a deeply personal endeavor. Fulfillment is not an end-state, but a state of being in which your daily life and activity is part of a larger sense of purpose. That purpose can be many things: being a better person; a better parent; a better provider; a better leader. Note that these examples are achievable and yet are not endpoints – you don't become a better person and then stop after you have "done it" and moved on. This is the essence of purpose, which is a state-of-being and not a finish line.

Determining your life purpose and prerequisites for personal fulfilment is far beyond the scope and subject of this book. This is one of those "authentic self" moments so many motivational speakers and writers talk about. It requires an honest look at your values and motivations outside of what external forces might place on you. This is also an exercise in recognizing and embracing your own mortality – every one of us will die someday. Being cognizant of that every day is a powerful avenue to recognize what is truly important in our lives.

One of my previous books, *Noble Echoes: Gifts from the Martial Arts for Modern Life*, delves into the subject of greater life purpose. It describes the lifestyle and ethical framework as conveyed by the traditional martial arts. Check it out if you are interested in discovering a powerful, productive way to live.

Defeating Procrastination

Procrastination is more than just a drag on productivity – it erodes the enrichment in our lives. A life worth living is one where goals and purpose are realized. This can only happen through action and effort – it is the essence of a participatory existence. Through action, we assert agency over our lives. Action is what creates the good in life. Action is what allows us to navigate hardship and the assorted unpleasantries of life. Without action, there is no quality of life or hope for the future. Procrastination is the enemy of action and thus an enemy of life.

As a universal feature of the human condition, procrastination affects us all. It is a self-defeating pattern of behavior that manifests from our innate defense and reward mechanisms. It is our ingrained compulsion to the avoidance of pain and preference for pleasure. Most of us would rather relax and indulge rather than spend the day working or exercising. As a fear response, procrastination has us avoiding the unpleasant aspects of work such as stress, difficult coworkers, or tedious tasks. It also functions to avoid psychological angsts like the fear of failure or judgement. These anxieties combined, with the readily available distractions, our brains automatically try to direct us away from the pain and toward the pleasure. How well we can counteract this aspect of human nature is a crucial exercise in self-control.

To procrastinate is to avoid doing something, and so the solution lies in getting started anyway. Understand that this is a battle with yourself and your human tendencies – this requires equal measures of sober self-discipline and self-compassion.

Reaffirming Motivation

The first step in overcoming the procrastination response is to remind yourself of your motivations – your goals, purpose, obligations, and mortality. This last part is important in that it fights fire with fire: the fear of our impending death and the limited time we have to accomplish our purposes often trumps the fear that creates procrastination.

Good Enough is Better than Never Done

Perfectionism is fundamentally unreasonable, and yet afflicts so many of us. Perfectionists place such high standards on every aspect of their work that they rarely complete goals in a timely manner – if at all. At its worst, perfectionism prevents us from starting on an endeavour in the first place. When you expect only perfection in your work, you are most often attaching higher importance to it than necessary. This creates an additional layer of anxiety around work which will trigger the fear response in your brain. Your brain, conditioned by eons of evolution, will do everything it can to create procrastination to avoid this stress. Thus, perfectionism goes farther than merely impeding the completion of tasks, it gets in the way of starting them in the first place.

Perfectionism can also prevent work from starting because it often forces us to wait for the "perfect" moment to begin: the perfect mood, perfect weather, perfect information, perfect timing, perfect energy, perfect inspiration, etc. As with everything else in life, nothing can be perfect. If you are in the habit of waiting for the "perfect" moment to begin a task, then you are allowing perfectionism to impede your ability to start.

Overcoming perfectionism requires recognition of its adverse effect on the achievability of larger goals. All goals are comprised of a series of smaller precursor objectives. Each one only needs to be "good enough" to progress to the next level. Anything beyond that is wasted energy and time – good enough and finished is better than never done. This does not mean that you should lower your standards of performance. Instead, it highlights the importance of being able to effectively set goals. Perfectionists have a hard time seeing what is good enough and end up wasting time fine-tuning results. The way past this starts with goal setting: before starting on any task big or small, take a sober look at what it is you are trying to accomplish. Define, in as specific terms as possible, what the minimum criteria for success are. From here, make it a point to stop working on your task the moment these criteria have been met. This extra mile of pre-planning allows you to set the "win" criteria of your work before you start. As you have preemptively decided what is good enough to finish your work, you have inoculated yourself against the allure of perfectionism. Make this a commitment to yourself ahead of time – when the end criteria are met, STOP. It's good enough, and that's good enough.

Getting Started With a Single Step

As a saying, "Just do it" is about as overplayed and droll as you can get. However, sometimes "just do it' really is all the advice you really need. In undertakings large and small, the first step is commonly the hardest one to take. It is the very thing that procrastination is trying to stop. This is because once you take the first step in working on a task you build inertia toward the next step and so forth. For this reason, the most tried-and-true approach to quickly defeating procrastination is to just put your head down and do it – even if all you are committing to is to do a single small step. In fact, by committing to only a single step, you are lowering the barriers in your brain to act. Remember, procrastination is born out of the fear response. To our brains, the bigger the volume, difficulty, unpleasantness, or mundanity of the workload, the bigger the fear. By reducing the commitment to a single step, the brain sees a smaller threat and so a more surmountable level of fear.

As you are attempting to hack past your brains' need to avoid stress, it helps to choose a step or action that is easy or quick to accomplish. You might also want to pre-determine a list of tasks that fit this criterion because by completing tasks, you will be building mental inertia to continue working. The more tasks you complete, the easier it will be to escalate to bigger and more difficult ones.

To this end, one of the more popular productivity rules comes from GTD: the two-minute rule. This rule holds that if any task is determined to take two minutes or less to accomplish, you do it without hesitation. In theory, this rule is intended to complete such tasks immediately as they come to you – avoiding capture into your task list. In practice, such short tasks will find their way into your task list anyway – just because a task is short does not mean it can be done the moment it comes to you. In any case, the two-minute rule is popular for a reason, and this is a great situation in which to apply it.

Breaking Down Big Tasks

Just about any undertaking can be broken down into shorter component tasks. You should already be in the practice of doing this when planning your projects or larger-scale work. If you find yourself procrastinating, consider breaking it down into smaller action steps. This strategy is much the same as the previous discussed above, except you are acting on larger-scale tasks rather than short standalone ones. Use this approach when you do not know where to begin on a complicated task – it creates a clearer path on what needs to be done to complete it in simpler terms. Whenever faced with complexity, the best course to take is to deconstruct it into simpler parts.

Most task managers solutions support the creation and organization of subtasks. Be sure to familiarize yourself with this functionality in your chosen tool.

Getting Past the Big Bad Stuff

At the opposite end of the above strategy is the practice of tackling the most parts of your workload first. Sometimes referred to as "**eating the frog**", the metaphor alludes to tasks that are the most distasteful or difficult on your task list. The strategy here is to complete these tasks first before anything else. The goal is to get them out of the way as soon as possible – paving the way for the rest of the day or session to be more pleasant and thus more conducive to productivity. This approach is ostensibly a mood hack intended to foster a happier mood and better-quality workday.

A similar strategy to the above identifies the tasks that comprise the essential must-do tasks of the day and do those before anything else. With this method, the most important work is front-loaded to the beginning of the day. As with eating the frog, the rest of the day, and thus your mood, is made more agreeable knowing that the most important and/or daunting work is already completed.

Though these strategies work for many people, you should be aware of some potential drawbacks. First, they are made with the assumption that you have more energy and focus at the beginning of the day (or session). This might not work for night owls and others with nontraditional activity cycles. Second, they can foster an exaggerated sense of accomplishment. In this case, the mind might figure that "enough" has been achieved for the day and loses motivation to remain productive and get additional work done.

Keep in mind that all techniques for overcoming procrastination are ultimately brain hacks. In using them, we are trying to work against innate human tendencies to avoid stress. We all possess different assortments of fears, experiences, tendencies, capabilities, and temperaments. The quest to defeat procrastination is one of self-discovery to find the methods that fit our unique psychologies.

Tomorrow is Never a Guarantee

Chapter 15: The Productive Mindset

Productivity is deliberate action, and action begins in the mind. Therefore, you must cultivate the right mindset. To be effective, you must have the right **attitude**, **motivation**, and **focus**. These are not easy traits to develop, as the mind is often far more difficult to cultivate than the body. However, without these traits, even the finest systems in the world will be ineffective. Your mindset is the basis for all productivity.

Productive Attitude

Your attitude is what sets the tone for your mindset – it is the bedrock upon which productive action is built. Many people let their current mood dictate their readiness to act. It is a tempting proposition: wait until the mood or inspiration strikes before getting started on achieving your objectives. After all, would you not accomplish more when you are good and ready? The flaw in this logic is that you will be at the mercy of your emotions and all the external factors that influence them. In a world of deadlines, in a life of finite length, there is no time to waste. You need to cultivate the disposition that gets you acting on your objectives regardless of negative inertia your feelings might generate. This is known as discipline.

A productive attitude is punctuated by a sense of discipline. **Discipline** is one of the most valuable life-skills anyone can have, and it is foundational to productive attitude. It is the strength of will to overcome hardship and/or emotional inertial to act. Discipline is not something that manifests via a single practice or trait – it is an assemblage of different qualities that must be cultivated individually. If you want to develop your sense of discipline you must create within yourself the right disposition.

The right attitude is built upon ambition, positivity, confidence, toughness, and integrity. **Ambition** is a goal-oriented outlook where you have a clear picture of a potential future you desire. Ambition is fed by a sense of **positivity**, which is simultaneously an acceptance of things as they are and a sense of optimism regarding things to come. This does not exclusively mean thinking that the future can only be good, but also that whatever may come, you will be okay. This aspect of positivity is an expression of **confidence** – the rooted belief that you can handle whatever life throws at you. Confidence also grants us the strength to set boundaries and embrace flexibility when things don't go to plan. Confidence is fostered through **toughness**, sometimes referred to as grit. Our culture incorrectly conflates toughness with dominance, bullying, and cruelty. This is not the case at all. Toughness is a sense of resilience – the ability to get back up after getting knocked down or to power through difficult situations. To be tough requires acceptance and surrender to the realities of life: that setbacks and failures inevitably happen and are not necessarily the end, and there is no shame in re-attempts. Toughness is a prerequisite of **integrity**, perhaps one of the most important virtues of all. Integrity is a deep-seated consistency in values and behavior, regardless of external factors. Its main feature is steadfast adherence to your convictions in the face of temptation or adversity.

Notice that the components of a productive attitude are complimentary. Each trait is a building block for the next:

Discipline is inspired by ambition.
Ambition built upon positivity.
Positivity build confidence.
Confidence fosters toughness.
Toughness enables integrity.

Productive Motivation

A key constituent of a productive mindset is motivation and the ability to muster it whenever it is needed. Motivation is the fuel that powers the meaningful aspects of our lives. This fuel must be sourced by the things in our lives that hold the most meaning. Motivation is built upon the greater purposes and goals we hold. As discussed earlier, discovering these aspirations and values is a deeply personal exercise. By doing the work to reveal your higher ambitions, you are laying the groundwork for your sense of motivation. It cannot be understated how important this is in creating a live worth living.

Even if you are very clear on your life goals and purpose, there will inevitably be times when motivation is in short supply. Even the most dedicated productivity gurus struggle with motivation. There is no shame here as it is part of the human experience. Even so, very few of us feel good about merely waiting for motivation to happen – the key to productivity is getting meaningful things done regardless.

What do you do when motivation is in short supply? Those of us lucky enough to have a deep well of discipline and grit have all they need to power through. The rest of us mortals will need a little more help, and this comes from the practice of affirmation. As banal and hokey as they may sound, affirmations are a popular and time-tested practice because they work. Think of affirmations as a personal pep-talk – the equivalent of a classical warrior beating a spear against their shield. Make this a regular practice whenever you find your motivation in short supply.

Summoning your motivation in this way can be done verbally or within your mind. This practice goes beyond reminding yourself of your goals and purpose. The trick is to foster a sense of urgency within yourself. And there are no better ways to do this than to contemplate your mortality. Life is finite and every day, the time we have in which to achieve our purpose is growing shorter every minute. The time we waste today will be gone forever. Think about your future self and how they might feel looking back at where you are right now. Honor your future self through this affirmation to find the motivation to act today.

Another thing to build your affirmations around are your obligations to others. Most of us have people who care about or depend upon us to achieve our goals and purpose. The most obvious people we have obligation to is our family – especially if you are a parent. We also have obligations to our friends, coworkers, and society. In many cases, fulfilling goals are tied to our obligations to them. Finally, you must honor the responsibility you have to yourself, particularly your future self. All these people are counting on you to act. Recognize this and be reminded of this whenever you need a dose of motivation.

Accountability is an important component of motivation. Inability or unwillingness to meet our obligations and responsibilities often create consequences. Being cognizant of these ramifications goes a long way toward mustering motivation to act. In some cases, consequences are obvious and felt directly – like failing to prepare for an important meeting. In other cases, such as missing a workout, the immediate consequences are much more subtle. These are instances where it helps to create additional accountability. For example, some people utilize the help of a friend to keep them committed to working out – if they fail to do a workout, they owe the friend money. Such an approach works to maintain motivation because it raises the stakes. The more severe the consequence, the more the brains' desire to avoid pain works in our favor.

In your efforts to foster motivation in these ways, be careful not to resort to self-deprecation. Beating yourself up for not being motivated does you no favors in the long-term. Remember, confidence is an important part of productive behavior. You must recognize that you are worthy of your aspirations if they are to have a realistic chance of coming to fruition. If you convince your mind that you are a piece of shit, then it will respond in turn by treating you as such. Self-compassion is part of a productive mindset.

Productive Focus

The digital age has introduced unprecedented avenues for distraction that propagates into every part of daily life. There is not a single moment in a typical day that cannot be infiltrated by a distraction from the internet. Even when away from our computers, smartphones and other connected devices are at the ready to steal our attention. Most of us begin our day on by checking our phones, often before even leaving bed. We live in a world of distraction that is never going away.

Focus is the final, indispensable component of a productive mindset. Without it, we are vulnerable to the allure of distractions all around us, be they digital or otherwise. Even in situations or environments where external disruptions are low, most of us have minds that will wander into distractions of our own making. No amount of planning or time blocking can address the tendency of our brains to meander attention away from the tasks at hand. To overcome this, we must cultivate the skills of mental focus.

Of course, some of us struggle with mental focus more than others due to psychological or physiological factors beyond our direct control. Conditions like ADHD, depression, anxiety disorders, and the like are real and have direct consequences on attention and focus. If you suffer from anything like this, then your first step in improving mental focus is seeking expert evaluation and treatment. There is no shame in having these conditions, nor in the pursuit of remedies. From here on out, I will be addressing matters of focus that lie outside of medical or psychological pathologies.

Mental focus is the ability to direct your mental efforts on a specific task, thought, or environmental element in a sustained manner. As a skill, mental focus is the practice of discipline. It is an ability that can be cultivated though repetition and practice. Mental focus entails more than merely filtering out distraction. It also involves the direction of your mental energy into the task at hand – utilizing every bit of your attention toward what you are doing.

Setting Boundaries

Focus is partly a function of setting **boundaries**. To this end, setting boundaries starts with the decision to be focused on the work at hand. This decision then becomes a proclamation: "I will be focused on this specific task for the next hour and will not allow anything to divert my attention." It is important to acknowledge your intentions regarding how your attention will be directed. This helps reinforce the agency you have over your attention span and the boundaries you set to protect it. This practice also affirms your culpability to yourself and your obligations – you are making a specific and conscious commitment to focus on the task at hand. It is far easier to deny distractions and irrelevant information after you have made a pledge to stay focused. Never underestimate the power of a promise – even to yourself.

Maintaining your commitment to focus means enforcing your boundaries by saying "no" to distractions. Much of this can be achieved preemptively. Before any focused-work sessions, you should establish your availability, or lack thereof, to any coworkers or family in your vicinity. Let them know that you will not be accessible during these times (emergencies notwithstanding). Such boundaries can be verbal or unspoken (like a closed door) but must be clear and direct. If others ignore these boundaries, you should correct them immediately – every time you make an exception, you are training both them and yourself that your time and boundaries are not important. The key to maintaining boundaries is consistency. Preserving these boundaries involves more than just other people – you must be just as assertive against yourself. For instance, if you catch yourself checking email or your social media when you should be working, act as though someone had burst through your door. Firmly, yet kindly, correct yourself and get back on task. We are often our own worst coworker.

Saying No to Context Switching

One of the greatest enemies of focus is **context-switching**, the redirection of attention from one unrelated task to another. For example: you are in the midst of a complicated spreadsheet and are interrupted by a slack message regarding a graphic file. In switching your attention from number-crunching to graphic design, your brain moves from one mode of thinking to another. Once the matter with the graphic is resolved and you come back to the spreadsheet, your brain must change modes back to number-crunching. Such scenarios are not the only ones that cause sudden switches in context. Within our own distractable minds, our thoughts often drift away from the task at hand – either into mere daydreams or unrelated work. In any case, task-switching is a common occurrence in the typical workday.

As many of us have experienced in moments like these, it takes time to recalibrate our thinking back to where we were before. This mental recalibration is not instantaneous. Even if it were, there is still the time wasted getting settled back into the task and find where you left off. If this type of attention redirection happens frequently during the day, these recalibration times add up. Multi-tasking does not work. The brain is simply incapable of directing meaningful attention to multiple tasks at once, and so multi-tasking is ultimately a practice of rapid context-switching. For every time attention is redirected, there is a refocus period that can add up quickly. Context-switching wastes time, and multi-tasking is a myth – don't do it.

Making Time

As had been said elsewhere in this book, time blocking is one of the most effective ways to get things done efficiently. Blocking time is a specific commitment to focus on a distinct type of work in a clearly defined timeframe. In many ways, time blocking is an expression of boundary setting: the time, commitment, and nature of our attention is clearly declared. As time blocks reside on your calendar, it means that you are arranging the rest of your schedule around this commitment. It is a powerful declaration of your intentions. This also makes it much easier to enforce your boundaries against intrusions.

Another way time blocking aids focus is that you can plan your blocks based on your work style and type of work context. You can set your time blocks for focused work into the hours where you are best able to maintain focus. For instance, if you are a morning person, you will block your most mentally intensive tasks or projects in the morning. Short-form tasks can be grouped by context or type (like spreadsheets or reporting) and blocked into the same timeframe. Known as "task batching", this is a more efficient way to use your time because it requires fewer mode switching between disparate types of work.

Time blocking is an effective foil to the menace of context-switching. It sets clear parameters on where attention is to be directed and sets boundaries to external interruptions. Through the practice of task batching, small tasks can be accomplished with optimal efficiency – it requires far less refocus time between tasks when they are of similar context or type.

Finally, time blocks can also be used for entire days (or weeks if you like) through day-theming. Just as with other blocking methods, setting aside full days for a specific focus allows you to set boundaries in your availability and primes your mind to stay in the necessary mindset for said work.

Maintaining Focus

It is rarely enough to schedule time to work and expect focus to automatically kick in when the time comes. Many of us struggle to get started on tasks, let alone on maintaining focus. How and when you begin working on a task or session sets the stage for your level of focus. Focus takes effort, and all effort requires energy. Different people have different times of the day in which their mental energy is maximized. This is known as the **biological prime time** and is why some people are early birds while others are night owls. If you know your biological prime time, then you know when you should be doing your most important and focused work.

Focus is also easier to maintain when performed in moderate sprints and not long sessions. It is far easier to maintain meaningful focus in short bursts. This is the basis for the pomodoro time management method and other similar techniques. By interspersing blocks of focused time with disciplined breaks, the mind is better able to keep focused for a longer period in aggregate. As with the body, it is easier to run a distance in shorter bursts than one long sprint.

The Power of Meditation

I would be remiss in a discussion about cultivating mental focus without mentioning meditation. It is the ultimate practice for mental development. Unfortunately, the rise of the so-called "mindfulness movement" has given meditation an undeserved reputation as some sort of panacea. Meditative practice absolutely enhances certain mental capacities. However, it is by no means a cure-all for any disease or disorder. In addition, there is nothing magical or metaphysical in its practice or mechanisms.

Meditation is a type of mental practice that cultivates the ability for the mind to create and maintain deep states of focus and alertness. It is a process of learning, and not unlike learning a musical instrument or other skill. Recall that the ability to concentrate on the task at hand is an essential productivity skill.

Meditation develops this very competence better than perhaps any other practice out there. The type of meditation described in this chapter is derived from the *Zen* tradition, itself the basis for the popular mindfulness methodologies.

A common misconception about meditation is that it creates a state of mind that is static and trance-like. This cannot be further from the truth. The meditative state is one of intense presence – a mind state free of internal distractions with all mental bandwidth available for the immediate moment. It is a state of hyperawareness and focus all at once. Additionally, it can be attained by anyone who puts in the requisite time to learn. Psychologically speaking, the state of mind meditation cultivates is known as the **flow state**, aka "the zone". This is the mental state of complete energized focus in which you are completely absorbed in what you are doing.

The Depths of Meditation

The process of meditation involves escalating stages. Each mental state must precede the previous one as they deepen. For the needs of productivity and organization, the first two stages are adequate. The first stage is known as the **undistracted mind**, and the second stage is the **focused mind**. If you are curious, the third meditative stage is the **alert mind**, which is utilized by the likes of martial artists and athletes. These stages of meditation are escalations of one another in that they create the requisite mental states for the next to exist in. Before you can have a focused mind, you must first have an undistracted mind, and so forth. One interesting thing about the meditative experience is how it is a cumulative process. Deeper levels of meditation are experiential and returning to them feel more akin to recalling a memory rather than a singular journey.

Meditative practice should be a regular occurrence. Sessions should take place daily if possible and last for at least fifteen minutes at a time. Session duration is quite important, and you should be working toward longer and longer periods as you progress –you are training your mind to maintain deep focus for extended periods of time. Beginners should learn to meditate while seated, either in a chair or in a kneeling pose.

The first state of meditation cultivates the ability of the mind to free itself from distraction and rumination This state of mind is often misinterpreted as being one free from thoughts – a suppression of the mind. This is not true, but rather a liberation of the mind from distracting thoughts. The mind will always be excreting thoughts. To be free of distraction by these thoughts, you must learn to ignore them. The way this is done is to focus your attention on something else, which in common practice is your own breath.

By clearing your mind and focusing only on your breathing, you provide the mind a target for its attention—making it easier to ignore and even eliminate your thought-stream. Maintain a slow and deliberate breathing process, drawn deep into the abdomen. Center the focus of your breathing on the abdomen, the muscles of the diaphragm. Count each breath, resetting the count each time your mind wanders into thought. See how high you can count before having to reset and always strive to beat your personal record. This is a slow and gradual process fraught with setbacks and plateaus just as with any exercise, physical or mental. In time, you will be able to maintain this mental state for increasingly longer periods.

After sufficient time meditating in this manner, you will notice your mind settling into ever-deeper levels of focus. This is where the second level of meditation begins to emerge. Subsequent meditative sessions will bring you further into this mental state. You will also notice that your meditations will bring you to these deeper levels more quickly. It is here where the capacity for your mind to focus is sharpened. In time, you will notice that your ability to focus on your work goes deeper than merely your attention span – it will encompass an acute commitment to your actions where every fiber of your attention and presence is centered on the task at hand

There is a misconception that meditative focus can only happen when the body is in a static position, and not useful when you need to move about. Not true. Martial artists employ this state as they perform complex and vigorous body motions. In combat, this mindset enables them to act with dynamic clarity in response to ever-changing tactical situations. This is the flow-state that athletes and masters of various arts of motion strive for. You too can learn to employ meditative focus as you move through the practice of moving meditation. This is an advanced practice, as you should have some proficiency at seated meditation first. **Moving meditation** is just what it sounds like – focusing the mind on the breath and detaching from thoughts while walking or running.

A Daily Practice

Experienced meditators learn that they can quickly call up the meditative mindset whenever needed. In having cultivated this ability, you will find that you can summon your ability to focus deeply at-will. Imagine how much more productive and organized you can be with such a superpower. It can be yours with sufficient practice. The ability for the mind to create and maintain deep focus is like a muscle learning to lift heavier weights for longer periods. The benefits of meditation result out of a cumulative effect of repeated practice – the more you do it, the better the results.

As with developing any skill, meditation is commonly fraught with setbacks. Our human minds are remarkably good at getting distracted – either by outside stimuli or thoughts of our own creation. One moment you are diligently monitoring your breath and suddenly you realize you were thinking about that Halloween episode of Night Rider. This will inevitably happen, as your "monkey mind" momentarily escapes your control. The counterpoise to this is not to become angry or discouraged, but to gently redirect your mind back on the breath and get back on track. Repeat. Repeat. Repeat. As time passes, the mind will go off course less often, but only with consistent practice.

The Mind is a Terrible Thing to Lose

Chapter 16: Productive Environment

The quality of your work environment plays a major role in your ability to focus and work effectively. Even the most unflappable among us are prone to distraction if the environment is disruptive enough. As is so often the case with productivity and organization, we are trying to accommodate the tendencies of our human brains. Our minds are distractable and temperamental – even innocuous aspects of a work environment can have significant effect on focus and efficiency. Lighting, background noise, ergonomics, temperature, and many other factors comprise the environment. Each one can have a positive or negative effect on how well we perform.

There are five components to a functional physical work environment: **Agency**, **location**, **comfort**, **organization**, and **ambiance**. When properly tuned, these elements work in symphony creating the best possible space for work to get done. Each component deserves special attention when optimizing your workspace. Few of us have complete control over every aspect of our environments. However, this does not negate the importance of addressing the parts we do have control over – optimizing one component can help negate shortcomings in the others. The quality of a work environment is most often a cumulative effect.

Agency

We all have a psychological need for a sense of agency – control over our circumstances. If you had not realized by now, cultivating your personal productivity and organization is a direct expression of this need. This human necessity is an important part of what we need in our workspace. The more influence we have over our environment, the better equipped we are to perform better.

Obviously, not everyone has the same level of control over their work environments. Some organizations allow more freedom in this regard than others. Certain types of work further limit how personalized a workspace can be. Even those who are self-employed or remote have varying constraints on what is in their control. Even if you have relatively little clout over your space, it is important to take advantage of what control you do have. Agency is too important a psychological need to give up completely.

An important aspect of workspace agency is the feeling that you are not an impostor, as if you are existing in a borrowed space. The most overt determinate of this is whether the space is permanent – a place just for you. Permanence instils a sense of stability which means lower stress which translates into better work. This is why you should abstain from working in spaces that serve dual purposes like the kitchen or couch – even if only to avoid the daily disruption of having to pack everything up.

Workspace agency is most overtly expressed through personalization. Decorations, photos, and other adornments are as vital to personal agency and expression as clothing. Make your workspace an expression of who you are – your values, motivations, and goals. Take as much ownership of your workspace as you can in this regard. However, be careful to not create a space that is full of distractions. Toys, posters, trinkets, and highly distracting items might add character to the environment but run counter to your productivity. As always, your efforts in exerting control over your work area is done in service of hacking your brain/psychology – productivity is always the end goal.

Location

Location is everything, as the adage goes. For workspaces, it can make or break your ability to work effectively. Your work locale sets the intrinsic level of distraction you must contend with. It also sets the tone for the ambient mood of your environment. If you are fortunate enough to have control over where you set up shop, it is essential to understand what to look for in a work location. For those of us who have less agency in this regard, this knowledge is useful for identifying strengths and shortcomings in the environment and compensate accordingly (if possible).

Where your workspace is located determines your degree of accessibility to others. Focused work cannot happen if you are frequently interrupted. Privacy is important in both mental and practical contexts. It is a basic psychological need and essential to the sense of personal agency and responsibility. From a utilitarian standpoint, it not only fosters the focus needed for good work, but also meets many of the legal or ethical requirements of certain work contexts – you do not want to be working on sensitive payroll information in front of an entire office, for instance.

On the flipside, you do not want your location to be so isolated that you are completely cut off from the world. Some work contexts require face-to-face interactions or even necessary interruptions. Many work dynamics also necessitate social interaction with others. Being a virtual hermit and hiding from the outside world does you no favors within these dynamics. The location and accessibility of your workspace must meet a balance between privacy and openness.

The best workspaces are also those that are near certain amenities. Unless you want to write novels in a shack in the woods, you obviously need to have access to a bathroom and snacks. However, the proximity of your workspace to amenities *does* have an impact on your productivity. For example, if you work far away from a restroom, you might have to walk past many potential distractions on your bathroom trips. This is to say nothing about the walk time it takes to get there – even brief trip times add up over the course of a single week. The closer you locate to such conveniences, the fewer diversions you might contend with. Conversely, amenities like restrooms, kitchens, and breakrooms are high-traffic locations, and working too close to them invites frequent interruption and distraction. As always, the trick here is to find the optimal solution – not too isolated and not too close.

Comfort

In both productivity and quality-of-life, comfort matters. Some people balk at this concept as being a sign of weakness or some other personal failure. Let them do so at their own productive peril. The mind performs at its best when the body is properly supported. Mental stress is exacerbated by physical discomfort. Drop the tough-guy (or gal) façade and do what is best for performance.

Ergonomics is the most overt place where comfort plays into productivity. Deep, focused work requires extended periods of time at the desk, bench, or workstation. Even conditioned bodies are prone to fatigue and strain. The more comfortably the body is situated, the longer it can perform without pain or exhaustion. A fatigued and sore body is not conducive to a focused mindset. Extended mental concentration exhausts the mind, and nagging physical discomfort will do nothing but exacerbate this. Never underestimate the mind-body connection.

Be careful not to confuse productive comfort for restorative comfort. When you are designing your workspace, the comfort you are seeking is that in which the body can maintain optimal posture for long periods. Posture in this parlance refers to body positions that encourage mental focus. Restorative comfort is the kind that is truly relaxing and conducive to rest or even sleep. Accordingly, your office chair will be one that you can comfortably sit up straight in for hours but not easily fall asleep in.

To address ergonomic comfort in the workspace is to evaluate furniture. Most knowledge workers sit in chairs all day, and so will need a quality chair that encourages proper sitting posture. Other jobs, and an increasing percentage of desk jobs, involve extended periods of time standing. In these instances, a padded floor mat can make this easier on the body.

Ergonomics go beyond chairs and standing mats. It extends to desks, keyboards, computer screens, and beyond. Your desk or worktable should be of sufficient height and size to best accommodate your chair and/or body. Computer screens should be positioned to minimize neck and eye strain. Keyboards and other computer input devices need to be positioned and formatted to curtail repeated muscle strains. Nearly every aspect of a workspace that you physically engage with is part of the ergonomics equation.

As our bodies age or work styles change, so do our ergonomic needs. If you find yourself with new body aches and pains, it is possible your ergonomic setup needs updating. This is not always because your previous setup was not working, but rather that your body has changed – it happens to everyone. Likewise, furniture and equipment wear out and need replacing, repair, or adjustments. In any case, ergonomics is an ongoing process.

Another aspect to a comfortable, effective workspace has to do with optimal climate. This is one of those cases where remote or self-employed workers are especially fortunate – few things foster office conflict more than agreeing on how hot or cold it should be! As with ergonomics, the right temperature of a work environment is not the most relaxing but the most conducive to productivity. This is a generally subjective affair, as different people thrive in different conditions. It also extends beyond merely the temperature of the air. For instance, I prefer to have a small fan blowing air in my face when I work, as it helps me stay focused (for reasons that are beyond me). Regardless, do not neglect the climate of your workspace when optimizing it for productivity – it really does make an impact.

Organization

As with digital spaces, the most productive work environments are those that are organized. Every tool, file, and piece of paper should have an unambiguous place where it belongs. Though some people swear by the efficacy of a cluttered office, it is rarely (if ever) a place of true efficiency. In addition, clutter and mess is unprofessional and can be a liability to your reputation. Treat your physical workspace with the same reverence and attention to organization as you would your digital life.

The digital age has reduced, but hardly eliminated, the usage of physical documents in most contexts. Most of us should still maintain a well-managed physical file system for such missives. Unless you are of the increasingly rare situation of relying on massive amounts of paper documents, I recommend a hybrid organization system. Documents that are either important or likely to be needed soon should be digitized. Smart phones are incredibly easy to use as scanners, and most notebook, storage, and document services provide apps to this effect. Your information repository service of choice should have such a scanning function in its phone app. If you have large amounts of paper documents for scanning, you might consider investing in a bulk document scanner. These scans should be stored and organized in your repository. This eliminates the need for complex file structures for your physical papers, as they are organized properly in your repository. In addition, this eliminates risk of loss of these files as the scans live in the cloud. So long as your digital scans are cataloged properly, you can store the physical versions with a low-hierarchy system: I personally store mine in folders by year only.

As with your digital files and emails, you should only retain physical files for a finite period. Also like their digital counterparts, different laws and practices apply to how long they should be retained. In some cases, certain types of documents would need to be destroyed. The same sources you would look to for advice on what retention rules you should apply for digital will likely also have advisements for your physical copies.

The key to a clutter-free space is that everything has a place. Every piece of technology or tool should have an easily accessible storage space. This includes peripherals, backup cables, travel cases, and everything else. Not only does this enable easier organization, but it also makes the process of cleaning and decluttering quicker. The less time you need to spend of cleanup, the better it is for your efficiency and peace of mind.

Finally, your workspace should be routinely cleaned. This means a daily cleanup ritual at the end of every work session. Not only does such a practice help maintain a neater office, but it serves a mental purpose – a ritual to train your brain to transition out of working mode. In addition, it is far easier to get started on work in the morning when the workspace is already cleaned and organized. Few things derail a day's work like having to cleanup from the night before. The end of every week should be punctuated by a deep-cleaning session to ensure the next week can start briskly.

There is a trend in the productivity and organization industries that elevates the concept of minimalization. The idea is that less is more and that the fewer things you have in your life, the less cluttered and "better" it can be. Such an ideology has led many to drastically reduce their workspaces to spartan levels of barrenness. If this is appealing to you and helps you simplify your work life, then do it. However, keep in mind that there is a middle way here too. You *can* have an intricate office layout and inventory *and* maintain harmony and organization. Organization and harmony are not opposites to complexity. Before you jump onto the minimalist bandwagon, take time to consider if what you seek is organization and not necessarily simplicity.

Ambiance

Ambiance is the "mood" of a workspace – the cumulative effect of environmental factors on our mental state. The quest for an optimal workspace is a quest for optimal ambiance. Every component of the work environment discussed thus far is part of it. You can think of productive ambiance in the same way as you would a romantic evening. Just as you would use lighting, color, sound, sent, and texture to make the room conducive to amorousness, you can do the same for productivity (do not try to do both at the same time). As with romance, people are receptive to different environment factors and preferences that influence their productivity. However, there are many commonalities here that are shared by more people than not, and so should be the starting points in creating a productive workspace.

Lighting

Much of the mental atmosphere of a space is determined by lighting and color. A work environment should provide ample lighting to be able to see what is going on, but not so glaring that it is hard to see. Some people prefer well-lit environments like most traditional offices have. Others do better in dimmer, lounge-style lighting. Similarly, the color scheme of the environment has effects on the mind and mood. While people might differ on preference for bright versus subdued color schemes, most would regard garish, high-contrast, or poorly matched coloration as jarring to concentration.

There has been significant research and conjecture about the effect different colors and temperatures of light have on mood. This has been bolstered by the booming popularity of variable LED "smart" lighting systems. Claims about how light colors influence our brains range from well-informed to outright pseudoscience. Some even go so far as to promote lighting as a panacea for virtually every mental or physical ailment. However, there *is* some good research that indicate real effects lighting has on the mind. Reputable studies have shown that blue lighting tends to promote calmness and lower heart rates while red hues tend to do the opposite[1]. The jury is still out on the extent of these effects and more so on how they influence productivity. Regardless, it is true that the mind is better able to focus in lower-stress environments. Though there is merit to some claims about red vs blue light, be skeptical of the more fanciful claims about ambient

---

[1] Ayash, Kane, Smith, Green-Armytage. 2015. *The influence of color on student emotion, heart rate, and performance in learning environments.* Color . Volume 41:2

light color. As stress and mood is so subjective, do not be afraid to experiment and see what you prefer.

Sound

Most people probably think about music and background audio when thinking about workplace ambiance. This is indicative of how influential environmental sound is to a productive workspace. One of the nice things about the internet age is that streaming services and headphones allow just about anyone to set their own soundscape for work. This makes the audio environment one of the few places most of us have total control over.

One thing that gets overlooked about a productive soundscape is the importance of silence. There is a well-known tendency amongst people driving cars who instinctively turn down the music volume when they need to increase concentration on the road. Research on this type of phenomena indicate that the brain, when focusing on something visual, will reduce the auditory information it receives. The brain is literally freeing up bandwidth to dedicate to what it is concentrating on. This tendency is what triggers the instinct to turn down the volume in situations like driving. A similar effect is seen when the mind focuses on auditory information – many people instinctively close their eyes to free up mental processing power.

Even the most ardent music lovers will have moments when silence is best - it strengthens the ability of the brain to concentrate more deeply when needed. Therefore, it is vital that your workspace can be silenced at-will. If you are fortunate to have your own office room, then being able to shut the door is usually sufficient. If you work in an open office or public space, attaining silence will be harder to do than adding audio. In these cases, you should have a contingency space to relocate to in high-stress situations in need silence. Noise-cancelling headsets exist, but bear in mind they do not completely filter out all background sounds.

The science on ambient music and productivity is just as varied as that on lighting. Just as with lighting, different people have different preferences for optimal working soundscapes. Conventional wisdom holds that the human brain has more difficulty focusing in noisy environments. This is true for many people, though hardly universal – many people get work done in coffee shops or commuter trains. Some people thrive in busy environments. Again, it falls to you to figure out what type of auditory background is best for your productivity.

If you do not know where to start in optimizing your soundscape, consider some of the following guidelines. First and foremost, avoid music or sounds that you find distracting – no matter how much you enjoy it. The sound profile of your workspace should serve your productivity and not as entertainment. Avoid utilizing music with lyrics. So-called "lo-fi" music is often effective in lowering stress and increasing concentration. This type of electronic music features softer sounds, consistent beats, no vocals, and yet still has a subdued energy. Regardless of your music usage, higher volume is detrimental to concentration. It should be just enough to drown out outside distraction, but not blaring.

Digital Ambiance

We all live and work on our computers, and so our devices have ambiances all their own. Even if you have a beautiful office with great lighting and sound, a hot mess of a computer environment will negatively affect productivity. Treat your digital environments with the same care and deference you would your physical spaces.

Just as with an office space, a computer should be organized with a place for every type of file and program. If you are following the guidelines presented in this book, then you have a good idea on how to effectuate proper digital file organization. However, there is more to fostering an efficient computer environment beyond file organization. Every device and operating system offer myriad places to access your files and apps: desktop icons, start menus, finders, dashboards, taskbars, docks, etc. These should all be organized and free of clutter. It also helps to keep access to distracting files and programs away from any quick-access menus or locations. For instance, your videogames should not have shortcuts on your desktop, dock, or taskbar – keep them out of sight to avoid temptation.

Just about every device lets you customize system color schemes and brightness. You can implement your lighting preferences here as well. More and more applications and operating systems offer dark modes and similar configurations. In many cases, you can even set times of day for when certain lighting settings are applied. If you are using external monitors, do not forget that you can configure lighting, color, and other settings on them directly as well.

Harmonious, but not Hypnotic

A common mistake people make when optimizing their work environment is that they make things too comfortable. It is easy to misconstrue a calming environment for a relaxing one. A relaxing space is one that is conducive to rest. If the lights are too low, the temperature too cozy, the seat too soft, and the music too soothing, you may end up encouraging sleep instead of productivity. A truly productive workspace is one that is calming in a way that reduces work stress and frees up mental resources for concentration. If you find yourself frequently on the brink of nodding off, consider that your work environment might be working against your energy levels.

As with every other part of organization and productivity, creating the optimal work environment is a perpetual process of evaluation and improvement. Office and computer spaces can get messy and cluttered over time. Familiar spaces, layouts, and ambiance can lead to feelings of boredom. Such mundanity can lead to a decrease in productivity and work engagement. In cases like this, which are not uncommon, the remedy is to reevaluate and refresh the workspace and make proactive changes.

Home is Where Your Stuff is – Work is How it Gets There

Chapter 17: Productive Body

It is easy to see why productivity and organization are seen as exclusively mental/psychological practices. After all, one does not go to the gym to improve their ability to manage tasks or label files. This is true in the literal sense but ignores the huge role our bodies play regarding our mental state. The mind-body connection is a very real thing, and hardly the dominion of new-age pseudoscience. The state of our bodies influence energy level, mood, endurance, and the ability to manage all varieties of stress. Being physically unhealthy places extra stress upon the mind. It's easier for mental stress to wear us down when our bodies are fatigued or distressed.

Physical health is an important part of productivity and all-around quality of life. Fortunately, you need not be a super-athlete to be healthy – just at or above the baseline threshold of wellness for you. This full breadth of this subject is far beyond the scope of this book. Any initiative to improve physical health must be undertaken with care and under the supervision of a doctor and other experts. It involves proper nutrition, right exercise, and the appropriate level of commitment and intensity. If you are out of shape or have lingering health issues, you owe it to yourself, your family, and your productivity to do something about it.

Sleep

The most immediate factor that dictates your physical and mental energy level is how much sleep you get. The peer pressure in modern work ethic has made lack of sleep into a something to be proud of. The less you sleep, the more you sacrifice for your job which somehow means you are working harder. This toxic mentality has equated exhaustion with success and has undoubtedly contributed to illness and even death. Ironically, the more fatigued you are, the less hard you are even capable of productive work.

Quality sleep is essential for productivity. Insufficient sleep impairs your ability to concentrate and stay on task. It even diminishes the capacity for the brain to solve problems, perform complex tasks, and even reasoning ability. If your work involves making weighty decisions, lack of sleep is a sure way to get things catastrophically wrong. Meager sleep also greatly impacts your mood and can make you irritable and short-tempered. It goes without saying that irritability is not a useful trait to have in any workplace – especially if you must contend with bosses, clients, or employees. Even the most pleasant and cordial person can be reduced to an angry, temper-prone mess given sufficient mental exhaustion. Consistently poor sleep also has direct negative consequences for the body – particularly for cardiovascular health.

The proper amount of sleep you should be getting most likely lies between what is recommended and what you prefer. Conventional wisdom posits that eight hours is the minimum quantity. As always, people are different and are refreshed by contrasting amounts of sleep. In addition, there are different theories and preferences for sleep patterns. Some people swear by the benefit of napping and shorter sleep periods. Others can only function with a single extended period of sleep. Further complicating things are the effects age, fitness, and myriad other factors play in optimal sleep.

As with lack of sleep, there is a cultural obsession with rising early as a purported mark of hard work and productivity. The thing is, not every person is a natural early bird, nor do they have the luxury to be able to retire in the early evening. Also, the rising proliferation of remote and asynchronous work is eroding the ubiquity of traditional working hours. Nevertheless, if you are more alert and energetic in the afternoon or even night, the world is increasingly accommodating to alternative active cycles. Night owls, rejoice.

Rest

Intermittent rest periods are important for optimal productivity. The body and the mind have limits to how much sustained activity and focus they can effectively maintain before performance drops. Even sitting at a desk for extended periods creates fatigue, which eventually impedes the ability of the mind to concentrate. The remedy for this is regular, disciplined rest periods throughout the day. Such breaks *must* involve stretching and moving the body. Just as such breaks limber up the body, they also limber up the mind – a chance to stop focusing and relax. The efficacy of this practice is why it is a big part of many time management techniques, like the pomodoro method.

Relaxation

Just as regular rest periods during workdays are important in maintaining productive output, so do longer periods of relaxation and restoration. This is another place where the toxicity of many work cultures creates problems – when taking time off for mental renewal is regarded as a sign of weakness and laziness. Vacations, breaks, and especially "mental health days" are taboos in these places. Do not accept this for a minute. Such ideology, rooted in an obsession with output, actually creates the very problems it seeks to avoid. Sustained periods of intense work stress and exertion without rest is dangerous and leads to burnout.

Burnout is a very real condition and is both unproductive and a serious threat both physical and mental health. It is a state of extreme emotional and mental exhaustion brought on by protracted periods of stress. Though the initial stressors are mental, they take a toll on the body resulting in lethargy and even physical pain. Though some jobs inflict more mental stress than others, even relatively mundane work can lead to burnout with enough intensity or duration. Just as with physical fatigue no one is immune from mental exhaustion, and thus there is no shame in succumbing to it.

Taking regular periods for extended relaxation is essential to avoid burnout and maintaining optimal overall productivity. Prolonged seven-day workweeks might look good for your reputation as a hard worker but will kill you and your output in the end. The harder you push your mind and body, the more drastic the reduction in efficacy and results.

Stay cognizant of your mental state and recognize when you are in a state of exhaustion or burnout. When this happens, the best thing you can do for your productivity is to make time for restoration. Burnout and general exhaustion cannot be remedied by short breaks or even half-days. At a minimum, it should entail a full day of disciplined relaxation. More acute cases of mental fatigue will require more than a single day and would benefit from a long weekend or even a full week of rest. It is not uncommon to feel twinges of guilt over taking time off as if you are betraying your responsibilities and obligations. This is a normal, though irrational, emotional reaction. In times like these, remind yourself that these extended breaks are specifically taken to recharge your mental energies. Renewing your mental acuity is part of meeting your responsibilities and obligations.

Be sure that the time you allot for relaxation is actually spent relaxing. This necessitates the practice of **disciplined rest**. Work has a knack for creeping into our personal time - especially in the age of smartphones. Countering this requires disconnection and boundary-setting. Your rejuvenating time must be free from interruptions and the temptation to do work. Inform your clients, coworkers, and family that you are on a break and will not respond to any non-emergency disruptions. If they ignore these rules, firmly hold your ground, and maintain your boundaries. These boundaries must also be maintained against yourself – resist the temptation to "check in" on work or do anything that is not relaxing or rejuvenating to you. Treat the time you set aside for relaxation as important to your work as any high-level meeting or deliverable.

Endurance

There is an old legend about the origins of the martial arts. An Indian prince named Daruma traveled to China in the 5th Century to introduce Zen Buddhism to the region. This variety of Buddhist practice is centered upon intensive meditation. It involves maintain deep levels of mental focus for extended periods of time. When Daruma reached the Shaolin Temple, he discovered that the monks were not capable of undertaking this type of mediation. They lacked the physical strength and stamina required to endure long periods of seated concentration. The weakness in their physical bodies impeded their ability to focus their minds. Daruma thus created a series of physical exercises for the monks to practice. Thus, the monks developed the physical rigor needed to endure Zen meditation. These exercises eventually became the basis for the martial arts the Shaolin Temple became famous for – supposedly the origin of all traditional martial arts.

As fanciful as this tale is, you can probably see the reason it is included here in this book – the ability of the mind to maintain focus is influenced by the conditioning of the body. As with meditation, productive work is done by focusing the mind on singular intentions. Additionally, this mental state must be maintained with the body in a controlled position: sitting at a desk, standing at a workstation, or even in motion. When the body starts to fatigue and lose its ability to maintain comfortable effective posture, mental focus and thus productivity decreases.

Productive physical endurance is an amalgamation of proper posture and body strength. Proper posture is essential to stamina whether sitting, standing, or in motion. When you carry your body poorly, you place unnecessary and excessive pressure on your body – most often the joints and muscles. Right posture is practiced with your weight held underside with your head held high over the shoulders. The objective of good posture is to keep the body in a state that requires the least amount of energy and strain to maintain position. This is where the ergonomics of your workspace also come into play – just make sure that you are utilizing it properly.

Posture is nothing without muscle strength to maintain it. Even the best posture can be physically exhausting if your muscles are too weak to hold up the body. The benefits of physical exercise are extensive and affect nearly every aspect of life in a positive way. At the very least, you should be doing thirty minutes of moderate exercise every day that utilizes all the muscles in your body.

Nutrition

What you put into your body affects your state of mind and thus your productivity. Many (perhaps most) people regard their mind as being separate from the body, when in fact our thoughts are the excretions of an organ – the brain. As with every other organ in our bodies, proper nourishment is required to keep it performing optimally.

Diet

There is a difference between food that advances overall health and that which optimizes productivity. A healthy diet is generally comprised of a balanced nutrient load and limited calories. However, even some "good" foods are not great for productivity. This happens when certain types of food adversely affect focus or energy levels upon consumption. Foods high in sugars or starches increase blood sugars rapidly, followed by a crash that leaves you feeling lethargic and hazy. Fatty foods and some proteins also diminish productivity in that they take more energy to digest, which drains from your overall reserve. In addition, many high-protein foods produce excessive gas in their digestive process, which can contribute to discomfort and impede mental focus.

Cultivating productive energy levels is largely a practice of managing blood sugar, and so the timing of when you eat is as important as what you eat. The types of foods that impede productivity, as described earlier, are best consumed after work. Eating too much or too little is also a consideration – you do not want to be racked with hunger all day but also want to avoid overeating, which leads to fatigue. For most people, the best approach are regular, small meals through the day – just enough to maintain energy levels but avoid overloading blood sugar causing a crash.

Hydration

Adequate hydration is essential to both physical and mental performance. As it is comprised of mostly water, inadequate hydration produces negative effects on the entire body. Dehydration reduces blood volume, which limits its ability to deliver oxygen and nutrients to the rest of the body. This puts extra strain on the heart, which must pump harder to compensate. The mental ramifications for low hydration are reduced energy, lower focus, and irritability. Maintaining hydration is essential to productivity.

Caffeine

As an addictive substance, caffeine is a cornerstone of many work cultures around the globe. So many people depend on it as their stimulant of choice that its use isn't going away anytime soon. Make no mistake, caffeine has a place in productivity so long as it is utilized with proper restraint. Caffeine indeed has positive effects in this regard. It stimulates the nervous system thus increasing focus, cognition, and alertness. While its stimulating effects are great for starting the workday or the occasional weary afternoon, overuse can adversely affect productivity and quality-of-life.

Consuming too much caffeine has adverse effects like any other drug. It is easy to ingest large amounts without realizing it – though caffeine is most closely associated with coffee, it is present in many varieties of tea and soft drinks. As a stimulant, excessive caffeine can make you jittery and irritable as it increases your heart rate. On the other side is the caffeine crash. This is the "come down" process of the drug. The results of this can include headaches, drowsiness, and inability to focus. To avoid these productivity-killing setbacks, monitor your caffeine intake and use in moderation. Additionally, avoid its usage in the evenings so it does not disrupt your sleep needs.

Say Your Vitamins and Eat Your Prayers

Chapter 18: Productive Habits

Without action, a system is nothing but a collection of ideas. Every system is ultimately an organized plan for action in achieving a specific outcome. Regardless how well they are designed, they can only create these outcomes through consistent action. No system of organization, productivity, hygiene, or fitness realize outcomes with a single one-and-done action. They require the application of routine, automatic, and repeating actions. This is the essence of what comprises a habit. Habits are the lifeblood of a system and what ultimately make them succeed.

Habits are comprised of the automatic behaviors you perform repeatedly. They are quite literally second nature – the tendencies in your attitude and behavior so deeply rooted into your personality that they are practically automatic. As so many of us are painfully aware, habits can be harmful as often as they can be good for us. Fortunately, habits can be changed, eliminated, and even created. This is not always an easy process, given the subconscious nature of habits, but it can be done with enough effort.

Managing Habits

The first step in altering, adding, or deleting habits is to understand how they work. The human mind is basically wired to seek pleasure and comfort and avoid pain and stress. This is especially apparent with bad habits – eating unhealthy food or avoiding exercise typically "feels" better than the alternatives. The good feelings and sensations our brain gets from such behavior stimulates the reward centers of the brain. When you binge on pizza, the resultant good feelings teach the brain that junk food is rewarding. Finally, every habit has a trigger that activates the brain to perform it. The mere presence of a pizza might be enough to trigger a habitual need to eat the whole thing. With something like exercise, it is anticipated discomfort that triggers the brain to habitually abstain. Thus, the habit is reinforced in the mind as a source of pleasure or an effective strategy to avoid pain. The more this happens, the stronger the habit lives in our subconscious.

Hacking Bad Habits

The key to hacking your habits is to understand why they exist. Since habits are perpetually repeating actions and must be assessed against their long-term effects over their short-term (emotional) benefit. Good or bad, your habits are defined by the outcomes they serve to produce. In nearly every case, these outcomes are either about avoiding or attaining something. If you want to determine what your existing habits are, look for these outcomes and the ingrained behaviors that produce them. For bad habits, this will provide insight into why your brain created them in the first place.

What is a bad habit? A bad habit is any automatic, repeating pattern of behavior that creates an undesired outcome – especially regarding long-term consequences. Skipping workouts or overindulging in alcohol feels good momentarily but create adverse effects after long-term repetition. Always look to the long-term ramifications of your habits.

In knowing what a habits' outcome is, you can readily identify the behavior that creates it. These behaviors are not as obvious and clear-cut as simply "do the thing that achieves the objective of the habit". You need to look for the nuance in these behaviors. Look for the patterns in these behaviors. Are they dependent on the presence of certain external factors? Can they only happen at certain times or places? Do you need to be in a certain type of mood or mindset for these behaviors to play out?

When you know what triggers your bad habits, you have already won half the battle against them. Some habits can be defeated simply by avoiding, removing, or negating the triggers.

However, most habits have multiple or unavoidable conditions that activate their triggers. In cases like this, you need to have a plan of action in place when these triggers fire – you are replacing what was once an automatic behavior with a conscious choice to do something different. You must have a new alternative behavior planned as a substitute. Since it must compete with the habitual pattern, this new behavior must be unambiguous, elegant, and as easy as (practically) possible. For example, if you are trying to break a habit of an afternoon energy drink, program a new behavior to grab a ginger ale the moment you feel the urge to reach for a Red Bull.

Changing habits takes resolve, discipline, and time. It is rarely an instantaneous process and is seldom easy. For this reason, you should only work on changing one habit at a time. The more behaviors you try to change at once, the less mental energy you will have available to devote to each one. Commit to the necessary time and practice self-compassion – you will have setbacks along the way, and that is okay so long as you stay the course.

Programming New Habits

Many of the same tactics for changing existing habits can be utilized in the creation of new ones. In most cases, creating new behaviors is easier than reprogramming old ones. The most difficult part in creating new habits is maintaining consistency. The stronger your resolve and sense of personal discipline, the better.

The first step in establishing a new habit is to clearly define the desired outcome. Habits are not one-and-done affairs, and so your outcomes will rely on repeating actions. For example: you want to program an exercise habit, and so the outcome will be "I exercise for one hour at least five times per week" rather than "I will get in shape". Your habits' outcome must define a specific repeating process that is clear, measurable, and attainable. Many people make the mistake of devising habits that are unreasonable in their attainability, which sets them up for inevitable failure. Keeping with the previous example, it is more reasonable to (initially) aim for a modest exercise regimen over "I will work out twice every day for three hours". Remember that you can always take established good habits and improve on them. But you cannot improve on a good habit until said habit has been programmed in the first place!

The second step to habit creation is to establish the specific actions and their respective triggers. Just as with defining the outcomes, actions must be clearly defined and attainable. "I exercise for one hour every morning before work" is specific and feasible while "I will exercise in the mornings" is not as much.

Every action must have at least one unambiguous trigger: "The first thing I do after waking up is get my bag and head for the gym." Most habits should have variations on their triggers to account for expected common variations: "I will grab my bag and head to the gym before I get ready for work". Habits are easier to program if the actions themselves are preestablished as well. For example, workout routines are easier to maintain when you know exactly what each session will entail. If you must improvise every workout session on-the-fly, it adds extra effort to the action that can further entice your mind to avoid it. Your habit actions should leave as little room for "winging it" as possible.

The final phase in programming a new habit is the execution. It is here that you should be maintaining a log of your progress – this can be done with something as simple as a spreadsheet or as complex as a dedicated fitness app. Doing so allows you to track performance and determine the efficacy of your efforts. The sooner you can identify that you are in a rut, the sooner you can adapt to overcome it. These logs should include more than just a checkbox on whether you did it or not, but also a record of your feelings and circumstances when you succeed or fail. Logging your habits also creates a visualization of consistency – especially in your streaks of success or failure. It does wonders for motivation when you can clearly see that you are on a multi-day streak of success. It also can instill a sense of urgency in cases of multi-day fail streaks. This is because tracking progress fosters the all-important sense of accountability to your commitment. You can further bolster the efficacy of accountability by inviting others to help you in your efforts to create habits. Share your habit goals with some safe and perhaps not-so-safe people. You will be amazed at how a little bit of shame can affect your motivation – just be sure not to overdo it.

Effective habit tracking will identify the obstacles and setbacks that impede development. Take regular inventory of these impediments and do not shy from being honest with yourself about why they happen. You are looking for explanations, not excuses. For consistent points of failure, you will need to make appropriate countermeasures to overcome them. For example, if you keep missing workouts due to oversleeping, you might need to improve your sleep/wake patterns or even move your workouts to later in the day.

The biggest hindrance our brains create in habit forming is their ability to find excuses and loopholes. Be cognizant of these and be sure to log them when your brain creates them. For instance, your mind might tell you that it's too tired to exercise or that the gloomy weather outside has killed the mood. Even if you succumb to these allurements, be sure to record them. You must contend with these self-sabotaging thoughts and feelings. Knowing what types of excuses your brain creates is half the battle in overcoming them. The better you understand what they are, the better equipped you are to overcome them ahead of time.

The Power of Routines

With so much in life that is uncontrollable, the power of routines cannot be understated. Routines are like a refuge in a world of chaos – an affirmation of the aspects of ourselves that lie under our control. Routines constitute the structure that underlies our everyday lives like the rhythm of a song. Routines infuse structure into your day, which boosts the psychological need for self-determination and personal agency. Maintaining routines also creates a grounding effect in the mind, even during periods of stress.

Like habits, routines can be leveraged to exert greater levels of productivity and organization in our lives. They can be thought of as aggregations of habits that are performed around specific times or contexts of the day. As a structuring practice, routines enable better and more efficient time usage. They automate redundant behaviors and actions, freeing up mental bandwidth for other tasks. Routines also create mental frameworks around our perception of time. For example, an afternoon routine that precedes lunchtime facilitates the switch of the mind away from "work mode" to "lunch mode". A similar application of routine can be made when initiating a specific type of activity, such as exercise or writing. For instance, if you perform a specific ritual before every workout, your mind (and thus body) will eventually learn to quickly switch into exercise mode at the onset of said routine.

Habits Are Like the Songs You Know by Heart

Conclusion

The true power of productivity is that it enables us to maximize the value of our time – especially in service of our greater goals and purpose. It maximizes the efficiency in which we meet our work and life obligations. This helps to maximize our available mental energies to commit to the truly meaningful aspects of life. Productivity is also a commitment to ourselves - an assertion over the agency we have over our lives. It is a powerful and deliberate way to live.

A Confluence

A system of productivity emerges out of the assemblage of our processes, tools, and behaviors. Our behavior dictates how we implement our processes via the tools at our disposal. No aspect of this trifecta can be ignored. True mastery of productivity can only emerge through mastery of each of these components.

To harness the power of productivity is to harness the power of the mind and its capacity to create and maintain focus. This is a skill that can be cultivated. To master productivity, it *must* be cultivated. There is no better method in fostering deep sustained focus than meditation.

A refined sense of productivity and its requisite skillset is one of the most valuable proficiencies you can have today. The technological marvels of our modern world imposed unprecedented levels of expectation on us all. By leveraging the power of technology, we can attain the type of superpowers that were recently the dominion of science fiction. We can already become cyborgs. Conversely, productivity means having a clear sense of our own priorities and purpose. In this way, we can be cyborgs but not mindless robots.

Understand that all systems are artificial, and no one approach will fit every need of every person. This is especially true of productivity, of which there are countless competing systems and philosophies. The best way to cope with this reality is to adapt the aspects of these systems that work for you and discard the rest.

The Time We Have

We all must die, and so the time we have is precious. Not everyone has the introspection or luxury to weigh the value they place on their limited time. Many people continue to live their lives with little consideration of their mortality or ultimate sense of purpose. Others are starkly aware of our mutual fleeting natures and see it as inspiration to make a greater impact with their time. If you read so far into this book, you are presumably of the latter type. Productivity is a life philosophy – the desire to make the most out of mortal existence and derive the maximum value and meaning out of it. Remember this whenever you question the merits of productivity.

As I stated at the beginning of this book, my intention with this work is not to convince you that my approach to productivity is best, but rather to instill a sense of awareness. I wish to foster a mindset that is cognizant of maximizing the value of limited time. When you are mindful of productivity, you are mindful of your mortality and the great importance of making the most out of life. It is an appreciation for the value of life and what you derive from it – a mindset not unlike the spirit of the warriors of old. Life is short. Make the most of it.

ABOUT THE AUTHOR

Gregory Shefler has worked in e-commerce for twenty years. He is a former senior leader and a co-founder at Coffee Bean Direct, an online-based coffee company. He has played nearly every role there is in the e-commerce field – requiring him to focus heavily on time management and organization. He has extensively researched methods of productivity to keep up with his professional, personal, and creative responsibilities. Gregory wrote this book as a way to share his productivity expertise with others.

Gregory has written two previous books. His first work is *Noble Echoes: Gifts from the Martial Arts for Modern Life*, an exploration of the ethical dimension of the martial arts. His other book is *Confluences of Online Marketing: A Holistic Guide*, which is an introduction to digital marketing for small businesses. Gregory can be reached at his website: gregoryshefler.com.

Made in the USA
Monee, IL
04 November 2026

67379676R00125